Powerful Ideas of Sc
and How to Teach Them

A bullet dropped and a bullet fired from a gun will reach the ground at the same time. Plants get the majority of their mass from the air around them, not the soil beneath them. A smartphone is made from more elements than you. Every day, science teachers get the opportunity to blow students' minds with counter-intuitive, crazy ideas like these. But getting students to understand and remember the science that explains these observations is complex. To help, this book explores how to plan and teach science lessons so that students and teachers are thinking about the right things – that is, the scientific ideas themselves. It introduces you to 13 powerful ideas of science that have the ability to transform how young people see themselves and the world around them.

Each chapter tells the story of one powerful idea and how to teach it alongside examples and non-examples from biology, chemistry and physics to show what great science teaching might look like and why. Drawing on evidence about how students learn from cognitive science and research from science education, the book takes you on a journey of how to plan and teach science lessons so students acquire scientific ideas in meaningful ways.

Emphasising the important relationship between curriculum, pedagogy and the subject itself, this exciting book will help you teach in a way that captivates and motivates students, allowing them to share in the delight and wonder of the explanatory power of science.

Jasper Green has worked in science education for over ten years as a teacher, head of science and most recently in initial teacher education. He is founder of thescienceteacher.co.uk and can be found on Twitter @sci_challenge.

'Jasper Green's book offers an antidote to the lack of vision which frames so many science curricula. Here you will find a fresh and innovative exploration of what it means to teach science. Drawing on much of the latest and best research and scholarship in education, it shows how these ideas can improve both the quality of student experience and their engagement with learning. The book reveals the nature of the complex challenge that it is to teach science. As such it should be essential reading for all teachers of science.'

– Jonathan Osborne,
Kamalachari Professor of Science Education,
Emeritus, Stanford University, US

'This is a beautifully structured book which explains some of the big ideas of science together with the best pedagogical strategies for teaching them. It is rich with examples, activities and practical applications of difficult concepts. And it is all about the challenge and the joy of making meaning, and how all our activities and tasks must work towards that goal.'

– Daisy Christodoulou MBE,
Director of Education,
No More Marking, UK

Powerful Ideas of Science and How to Teach Them

Jasper Green

Routledge
Taylor & Francis Group

LONDON AND NEW YORK

First published 2021
by Routledge
2 Park Square, Milton Park, Abingdon, Oxon OX14 4RN

and by Routledge
52 Vanderbilt Avenue, New York, NY 10017

Routledge is an imprint of the Taylor & Francis Group, an informa business

British Library Cataloguing-in-Publication Data
A catalogue record for this book is available from the British Library

Library of Congress Cataloging-in-Publication Data

Names: Green, Jasper, author. | Routledge (Firm)
Title: Powerful ideas of science and how to teach them /
Jasper Green.
Identifiers: LCCN 2020009065 | ISBN 9780367188658 (Hardback) |
ISBN 9780367188689 (Paperback) | ISBN 9780429198922 (eBook)
Subjects: LCSH: Science--Study and teaching.
Classification: LCC Q181 .G764 2020 | DDC 507.1–dc23
LC record available at https://lccn.loc.gov/2020009065

ISBN: 978-0-367-18865-8 (hbk)
ISBN: 978-0-367-18868-9 (pbk)
ISBN: 978-0-429-19892-2 (ebk)

Typeset in Palatino LT Std
by Cenveo® Publisher Services

Contents

Figures

Tables

About the author

Jasper Green has worked in science education for over ten years as a secondary teacher, head of science, and science lead for Ark Schools. Most recently he has been working on initial teacher education and MA programmes at the UCL Institute of Education.

Jasper studied biological sciences at the University of Oxford and later gained his PhD from the University of York. He has an MA in science education from King's College London which, together with his passion for science, social justice and teacher education, prompted him to write this book. Jasper is found on Twitter @sci_challenge and is also founder of thescienceteacher.co.uk – a website providing science teachers with free resources and ideas for the classroom.

Acknowledgements

Writing, whether it's an assignment, dissertation or book involves borrowing ideas and creating new ones, neither of which happens in a vacuum. I, therefore, have a number of people to thank who have made this book possible, either directly through giving feedback or indirectly through shaping my understanding of what it means to teach science well.

I want to start by thanking Frances Myers. She first showed a dyslexic 13-year-old, placed in a bottom set for science, the joy that learning biology brings. But perhaps more importantly than that, she showed me the positive difference teachers make.

Working in education, I have worked with many passionate and talented colleagues. They have been generous enough to share their expertise with me and cared enough to debate ideas when we disagreed – thank you. Of course, teachers are nothing without their students and so thanks to you all, young and younger, who have made the job of teaching science so rewarding.

Annamarie Kino at Routledge first approached me to write a book. I am grateful to her for allowing and supporting me to write *this* book and to Apoorva Mathur for seeing it through to publication. Dr Ralph Levinson not only did his best to teach me how to teach science when I was a beginning teacher, he also kindly agreed to give feedback on Chapters 1–6. Ralph's feedback was incredibly formative, but it was also written in a style that made the painful process of redrafting almost pleasurable. Jill White has played an important role in shaping my understanding of teaching in general, but more specifically on how to approach the notoriously complex task of planning science lessons. I am grateful for her feedback on Chapters 7–9. Professor Michael Young provided some wise words on the final draft that, fingers crossed, mean the aims of this book are now clearer. And I am grateful to Professor Jonathan Osborne for his feedback that, amongst other things, reminded me of the importance of conversation when learning new ideas.

Roni Malek kindly helped a biologist and a 'pseudo-chemist' get to grips with the wonders of physics by providing feedback on relevant chapters. It is Roni's demonstration in Chapter 11 that I've borrowed to introduce you to falling objects. Other valuable suggestions and ideas have come from George Duoblys, Philippa Franks, Thomas Kitwood, Dr David Paterson, Susie Sell, Charles Tracy and Dr Ruth Wheeldon. Of course, any mistakes that remain in this book are due to my inability to notice them.

And last, but by no means least, thank you to friends and family who have supported and encouraged me, always making sure I laughed along the way. This book is dedicated to you.

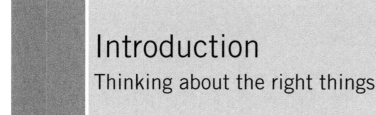

Introduction
Thinking about the right things

It was a blazingly hot afternoon and I was waiting apprehensively in my lab coat for Year 7 to arrive. 'Today we are learning about energy and food'. After fielding numerous questions about 'are we doing a practical today, sir?' and 'why are you dressed as a doctor?' we were ready to begin. Students busily completed a worksheet matching up key words before proudly declaring they were 'finished!' Unable to think of a suitable stretch question to keep them busy, I swiftly moved on to explain the experiment. Students made a dash for the apparatus and began setting fire to various snacks which were then used to warm boiling tubes containing too much water. As Pringles were plunged into the roaring Bunsen flame, hot oil started dripping down their heating needles causing a serious risk of burning, not to mention dissipation of energy. Black smoke filled the air as smouldering crisps lay abandoned on the bench. 'That stinks!' coughed one student. With a glance to the back of the room, even my university tutor had stopped writing.

Looking back at this lesson, I made a number of mistakes, such as a failure to establish routines and relationships and activity-focused planning with no thought to what students were actually doing. Fundamentally, this lesson failed because the students and I were thinking about the wrong things. That is, students were thinking about eating Pringles, dripping oil and who could finish the task first, and I was thinking about managing behaviour, distributing equipment and not setting off the fire alarm.

This book seeks to address some of these mistakes by exploring how to plan and teach science lessons so that students and teachers are thinking about the right things – that is, the scientific ideas themselves. Scientific ideas are powerful ideas because they give young people powerful ways of seeing the world that are dramatically different from their everyday experiences. Plants get their food from the air around them, not the soil beneath them; objects move even when no force is acting and mass is always conserved when paper burns.

But this scientific perspective involves more than just a knowledge of the facts of science (the substantive knowledge), it also requires young people to understand how these scientific ideas became established in the first place. It is through understanding the relationship between the scientific ideas and the nature of science that students come to appreciate science as a discipline of inquiry, which enables them to ask and answer questions of the world such as where humans came from, why there isn't yet a cure for cancer and why coldness doesn't exist. Indeed, it is by thinking about how and why scientific knowledge becomes established that students come to see its limitations. That is, scientific knowledge is tentative, subjective and driven, in part, by progress in technology and human ambition that has both positive and negative consequences for society.

But therein lies the challenge, as it is because scientific ideas and practices are radically different from intuitive, everyday thinking that they can be hard to learn and tricky (but wonderful) to teach. This means that if we are not careful, only some young people will have access to this knowledge and even fewer will understand its significance to their own lives. Science teachers then must become experts in what they do, so they can help *all* students to acquire these abstract ideas in meaningful ways.

At the heart of this teacher expertise lies deep subject knowledge, and so each chapter provides the opportunity to develop an understanding of one scientific idea and how to teach it. Scientific ideas are introduced, one chapter at a time, in a way that will support progression in your own understanding of biology, chemistry and physics as you move through this book. Alongside the subject knowledge sit ideas of how to teach it. These ideas are based on my own experience, working in schools and in initial teacher education, but the greatest source of inspiration has come from watching other teachers teach and seeing how their students learnt. Where possible, I have linked strategies to research so that you can better understand the rationale as to why these ideas may, or may not, work in your classroom. By using research in this way, I hope to convey some of the intellectual vibrancy that teaching science brings and encourage you to engage with these and other ideas as you develop your own philosophy of what it means to teach science well. You can find further ideas and resources, shared by many teachers, on my website – thescienceteacher.co.uk.

Whilst I am not teaching in classrooms at the moment, I have chosen to refer to 'we' throughout the book to create a sense that *we* are on a journey together. Please forgive me if it sounds disingenuous at times, but I hope the overall style makes for an easier read.

There are five sections to this book that you will encounter along the way. Section 1 looks at some of the aims of science education. In Chapter 1, we move beyond seeing science from a functional perspective, that is, for creating more scientists or knowing how to wire plugs, to seeing science as a body of specialised knowledge and a tradition of inquiry that can transform how young people see themselves and the world around them. Then, by introducing you to 13 powerful ideas of science in Chapter 2, I hope to frame science as a discipline of related, but distinct, ideas that provide the substance through which young people can ask and answer interesting and important questions that matter to them. Whilst each powerful idea in this book is a substantive idea, such as evolution by natural selection or gravity being a force of attraction between all objects, running through each of these ideas sit many opportunities to explore the nature of science. Powerful ideas then provide an education into both the substantive *and* disciplinary ideas of science and to do one without the other is to misrepresent what science is.

Section 2 considers how teachers can help students to acquire scientific ideas in a meaningful way and so requires us to recognise that knowledge is more than just information. We look at some of the distinguishing features of scientific knowledge that make learning science tricky in the first place, before exploring in more detail how we learn, drawing on ideas from cognitive science. But learning science (and doing science) involves emotion too, and so we explore some of the affective aspects that play important roles in initiating and sustaining learning in the science classroom.

Taking what we know about how and why students learn science, Section 3 looks at how to approach the complicated business of planning science lessons by thinking about the *'what'*, the *'how'* and the *'form'* of lesson planning. The *what* identifies what we want students to learn and do. The *how* takes the *what* and transforms it into something students can experience, discuss and think about. This involves thinking about the *best* ways to represent and portray ideas and so involves an appreciation of the important relationship between curriculum, pedagogy and the subject itself.

Then, taking the *what* and the *how*, we consider the form of lesson planning that introduces you to the six phases of instruction. The *form* looks at ways to introduce scientific ideas without overloading students by building upon prior knowledge and focusing attention on the important bits using carefully designed demonstrations. By allowing time for practice and application, students then refine their scientific understanding and start to integrate this scientific perspective into their lives so they can see the wonder in the everyday. These six phases of instruction are explored in Section 4 in more detail using a number of examples and non-examples so that we can consider what works (and doesn't work) in the classroom and why. Finally, because students won't always learn what we intended them to, Section 5 explores what responsive science teaching looks like so that we can respond and try again.

So, hold on tight as we begin our journey into the powerful ideas of science and how to teach them.

Section 1

Aims for school science education
For whom and for what?

Fallacies of science education

At 8:15 am on 6 August 1945, the first atomic bomb used against human beings exploded over the Japanese city of Hiroshima. The bomb, packed full of uranium-235, exploded 600 metres above the city as two pieces of uranium were catapulted into each other, creating a super-critical mass of radioactive material that initiated a chain reaction. Vast amounts of energy were released as uranium-235 atoms split apart by nuclear fission, creating temperatures on the ground that fused metal and melted glass. By the end of 1945, 140,000 people had died (Figure 1.1), but many more would go on to endure the horrific consequences of radiation poisoning. Whilst the atomic bomb was a weapon of war, it was also a product of science.

This uncomfortable truth, that scientific knowledge has the propensity for both good and bad, is rarely captured in the school science classroom. Instead, science is portrayed as a method in search of objective truths, without much thought as to why these ideas arose in the first place. But science is a human endeavour that takes place in a context and culture that influences what gets measured as these four examples illustrate:

- the Manhattan Project, the three-year, American-led effort to develop the atomic bomb, was only initiated in fear of Germany getting there first;
- Marie Curie, the first woman to receive a Noble prize, named the element she discovered polonium after her homeland country of Poland;
- in 2012, research funding in the United Kingdom equated to £241 per person with cancer, £73 per person with coronary heart disease and just £48 per person with stroke (Luengo-Fernandez *et al.*, 2015) and

Figure 1.1 The tricycle of Shinichi Tetsutani (then 3 years and 11 months) who died on 6 August 1945

Note: This tricycle was initially buried with Shinichi at home. Forty years later, Shinichi's remains were transferred to the family grave and the tricycle was donated to the Peace Memorial Museum in Hiroshima by Shinichi's father, Nobuo Tetsutani. Photo donated by TETSUTANI Nobuo to the Hiroshima Peace Memorial Museum, Curatorial Division 1-2 Nakajima-cho, Naka-ku, Hiroshima 730-0811, Japan.

- after the death of a 'star' scientist there is often a surge in publications from outsiders who resisted publication whilst established scientists were still alive. Often these publications advanced the field considerably by offering new directions (Azoulay *et al.*, 2019).

Whilst there are some very good reasons why we may want to avoid too many distracting stories when teaching scientific ideas (Harp and Mayer, 1998), there is a risk that portraying science in an overly sanitised way disconnects school science from the complexities of the real world, meaning that for many students science is seen as 'important but not for me' (Jenkins and Nelson, 2005). How much then of this messiness should we share when we teach science? The answer depends on what and for whom an education in science is for. Once we've sorted this out, we can decide which scientific ideas students should learn about and, importantly, how they should learn them.

Science for scientists

For a long time now, school science has been justified on the basis that we need to educate future scientists. This virtuous aim makes perfect sense until we consider that only 20% of the UK workforce need scientific training to do their jobs (The Royal Society, 2014). And, even if we are able to increase the number of students who want to work in science, there are genuine questions as to whether more scientists are really needed for this science pipeline; supply seems to be meeting demand in the United Kingdom and United States at least (Smith, 2017). So, whilst there is more work to be done to make sure this pipeline adequately represents the society science serves (DeWitt and Archer, 2015; Smith and White, 2019), it seems entirely undemocratic to justify science for all on the basis of creating future scientists. It seems then that we need a more equitable reason for why science is taught at school, one that is relevant for all students and not just a minority.

Utilitarian aims: useful or practical, rather than attractive

Seeing as not everyone can or wants to become a scientist, perhaps we should be looking for a more pragmatic and practical purpose for science education? Take, for example, knowing how to wire a plug or knowing the location and names of major human organs; surely these are valuable aims for everyone? I'm not so sure. Arguing for a science education for all based on the necessity to wire plugs or identify locations of important organs is problematic for a number of reasons.

For the most part, advances in modern technology have superseded many of these historically useful roles for science. A quick Google search identifies where our clavicle is located and most appliances now come with plugs attached. Then there's the question of whether scientific ideas are really that useful in the first place. The National Science Board of America (2018) reports that 27% of Americans think the Sun goes around the Earth, but this doesn't stop them from distinguishing night from day (in Europe the figure was 44% so Europeans don't go feeling smug!). Indeed, many of the misconceptions that make science teachers shudder, such as closing the door to keep the cold out, are actually very functional in our day-to-day lives. I would suggest that asking someone to close the door to prevent the transfer of thermal energy is unlikely to get you very far, unless, that is, you are in a room filled with scientists!

▌The scientific habit of mind

Maybe then it's not about such practical aims, but rather being able to 'think like a scientist'? Over 100 years ago now, the educationist and philosopher John Dewey (1910) gave an address to the 1909 American Association for the Advancement of Science annual conference at the Massachusetts Institute of Technology. Here, Dewey argued that it is the method of science that should be the primary outcome of a science education so that people can appreciate the 'scientific habit of mind' (p. 126) as opposed to focusing on 'ready-made knowledge' (p. 124). For Dewey, the significance of science lay not behind the content but behind it being 'an effective method of inquiry into any subject-matter' (p. 124).

The problem, though, with seeing science as 'a method' is that it divorces the process of doing science from the process of thinking about scientific ideas. In separating out the method from the ideas, it inadvertently undermines the role of scientific knowledge in the inquiry process. Let's take, for example, an inquiry into the factors that affect photosynthesis – the process by which plants produce their own food using light, carbon dioxide and water (amongst other things). You can't begin to design or carry out this inquiry unless you have sufficient knowledge of (a) the variables likely to affect the rate at which plants photosynthesise, (b) know how you will measure the effect of changing these variables and (c) know how to make meaningful conclusions from the results you collect. Ask an inorganic chemist to inquire about the Qin Dynasty – the first dynasty of Imperial China – and you won't get very far; not because she doesn't know how to inquire, but because she doesn't have the necessary knowledge of Chinese history to inquire with.

The second problem with seeing science as a method is whose method are we learning? Take the work done by an evolutionary biologist and a primatologist such as Jane Goodall. The evolutionary biologist may never perform an experiment in the traditional sense with glassware, goggles, solutions and measuring/changing variables. Instead they may spend hours at a computer modelling the evolutionary history of DNA sequences. The primatologist, on the other hand, may spend weeks painstakingly recording the behaviour of chimpanzees in the Gombe National Park in Tanzania. The archetypal model of a scientist making hypotheses, controlling, manipulating and measuring variables, whilst not wrong, is insufficient to describe the many different ways scientists work to generate new knowledge. Jonathan Osborne (2002) describes this as one of the many 'fallacies of science education'. What's unique to science then is not the process of how scientists generate knowledge – historians generate hypotheses, geographers make observations and cooks control variables. Instead, what is unique to science is a collection of scientific ideas that scientists use to inquire with and make sense of the natural world.

Scientific ideas as powerful knowledge

Michael Young, an educational sociologist working at the UCL Institute of Education in London, refers to these scientific ideas, amongst other academic knowledge students acquire at school, as 'powerful knowledge' (Young, 2009). Powerful knowledge is specialised knowledge, created by communities of subject specialists, that gives us the ability to think about, and do things, that otherwise we couldn't. Young calls this powerful knowledge, not because it exerts 'power over' individuals, but because it gives 'power to' individuals, taking students beyond their everyday experiences by allowing them 'to think the unthinkable and the not yet thought' (Bernstein, 2000, p. 30). For Young, schools are places where students should acquire this powerful knowledge.

To illustrate what powerful knowledge looks like in science, let's take the scientific idea that *'the diversity of organisms, living and extinct, is the result of evolution by natural selection'*. This is a powerful idea because it takes us beyond our everyday context and involves specialised knowledge of inheritance, genes, variation and natural selection that has been created by specialist communities of geneticists and evolutionary biologists. The idea enables us to appreciate our incredibility through our 3.5-billion-year history, yet, at the same time, recognise our fragility as no species will survive forever. We can recognise why there are flaws in our own anatomy; after all, we are built on compromise and not design. Being able to simultaneously talk, breathe and swallow comes at a cost that occasionally requires the Heimlich manoeuvre. Not only does science explain how we evolved, it also explains how millions of other species evolved, allowing us to appreciate their similarly impressive struggle for survival. With this knowledge of evolution comes an ability to inquire, to ask questions and to be curious: Why, for example, do men have nipples? Why is blue such a rare colour in nature and why do penguins huddle?

Understandings like these may seem a little extravagant as an aim for compulsory science education lasting more than ten years, but they are anything but. Just as a life devoid of reading or music would be less rich, a life devoid of scientific ideas is restricted only to what can be seen, heard, tasted and touched. There would be no words or ideas to describe genes, organisms, atoms or matter. We would not be able to explain why children look like their parents or why lightning is seen before thunder is heard. The world would be significantly less interesting and less meaningful. Puddles would simply soak into the ground, the Moon would be bigger than the Sun and global pandemics would be the result of witchcraft or 5G networks.

Scientific ideas, then, allow us to appreciate the wonder of the natural world (Hadzigeorgiou, 2012) by giving us access to an invisible world that provides explanations for the things we can see whilst also offering up new ways of thinking about the things we can't:

- when you dissolve salt in cold water you break the same ionic bonds as melting salt, but melting salt requires temperatures in excess of 801 °C;
- the Eiffel Tower can be 15 cm taller in the summer and
- the spider-tailed horned viper is so named because it has evolved a spider-like tail to lure in unsuspecting insectivorous birds.

Distilling down the aims of science education

The primary implications, then, for not knowing science are not practical or economic, they are emotional and concern social justice. All young people should have the ability to inquire about the natural world from a scientific perspective, just as they should learn about numbers, history, words and art. Not only does this scientific perspective give our everyday experiences meaning by transforming the bland everyday into moments of wonder, it also helps us to see how our actions have implications for others. Some people may see this second aim as civic action. But civic action makes it sound as if science education has been hoodwinked into serving another purpose, perhaps crossing over into the murky waters of moral education.

However, thinking how scientific processes connect humans is integral to the discipline itself; it requires no separate aim to be bolted on. It makes no sense to teach combustion without allowing students to see how the products of this chemical reaction have important implications for communities living at sea level. Neither should we teach about metal extraction without students understanding the environmental consequences of using too much kitchen foil. What students choose to do with this knowledge is of course their business, we just need to make sure this scientific knowledge and special way of seeing the world is available to all.

Summary

As we distil down the aims of science education, we need to be aware of how our own experiences can influence and bias what we think is important. For a science curriculum to have value, it needs to have value for *all* students, not just the few that will go on to become scientists or teach science. Well-meaning attempts to address this challenge by devising separate curriculums for scientists and consumers of science (sometimes called citizens) will inadvertently disadvantage those students who are already disadvantaged. Instead, a single science curriculum *for all* should focus on a set of core powerful ideas that are unique to science and honestly explore how they came to be. Through acquiring these scientific ideas, students have the opportunity to inquire freely about the world from a scientific perspective. Science education then has an important role to play in social justice in ensuring that all students, not just those who attend 'elite' schools, have the opportunity to acquire these elite ideas. It's time to take a closer look at what these powerful ideas of science could be.

Bibliography

Azoulay, P., Fons-Rosen, C. and Graff Zivin, J. S. 2019. Does science advance one funeral at a time? *American Economic Review*, 109(8), pp. 2889–2920.

Bernstein, B. 2000. *Pedagogy, symbolic control, and identity: Theory, research, critique* (Vol. 5). London: Rowman & Littlefield.

Dewey, J. 1910. Science as subject-matter and as method. *Science*, 31(787), pp. 121–127.

DeWitt, J. and Archer, L. 2015. Who aspires to a science career? A comparison of survey responses from primary and secondary school students. *International Journal of Science Education*, 37(13), pp. 2170–2192.

Hadzigeorgiou, Y. P. 2012. Fostering a sense of wonder in the science classroom. *Research in Science Education*, 42(5), pp. 985–1005.

Harp, S. F. and Mayer, R. E. 1998. How seductive details do their damage: A theory of cognitive interest in science learning. *Journal of Educational Psychology*, 90(3), pp. 414–434.

Jenkins, E. W. and Nelson, N. W. 2005. Important but not for me: Students' attitudes towards secondary school science in England. *Research in Science & Technological Education*, 23(1), pp. 41–57.

Luengo-Fernandez, R., Leal, J. and Gray, A. 2015. UK research spend in 2008 and 2012: Comparing stroke, cancer, coronary heart disease and dementia. *BMJ Open*, 5(4), e006648.

National Science Board. 2018. *Science and engineering indicators 2018*. NSB-2018-1. Alexandria, VA: National Science Foundation. Available at: https://www.nsf.gov/statistics/indicators/. [Accessed: 15 December 2019.]

Osborne, J. F. 2002. The challenges for science. Education for the twenty-first century pontifical academy of sciences, Scripta Varia 104, Vatican City. Available at www.pas.va/content/dam/accademia/pdf/sv104/sv104-osborne.pdf. [Accessed: 20 December 2019.]

Smith, E. 2017. Shortage or surplus? A long-term perspective on the supply of scientists and engineers in the USA and the UK. *Review of Education*, 5(2), pp. 171–199.

Smith, E. and White, P. 2019. Where do all the STEM graduates go? Higher education, the labour market and career trajectories in the UK. *Journal of Science Education and Technology*, 28(1), pp. 26–40.

The Royal Society. 2014. *A picture of the UK scientific workforce*. Available at: https://royalsociety.org/topics-policy/diversity-in-science/uk-scientific-workforce-report/. [Accessed: 15 December 2019.]

Young, M. 2009. Education, globalisation and the 'voice of knowledge'. *Journal of Education and Work*, 22(3), pp. 193–204.

Powerful ideas of science

▌Powerful idea focus

- *The cell is the basic structural and functional unit of life from which organisms emerge.*

When Ralph Levinson was 16, he had an epiphany – in a secular sense.

> One afternoon when I was not focusing on anything in particular it struck me, as if lightning had flashed out of an unlikely sky, that the ideas of atomic chemistry were awe-inspiring. Here was a world of atoms, protons, electrons, spin, orbitals which explained so much; non-perceptible, real yet entirely the product of human thought and social practise.
>
> *Taken from Levinson (2018, p. 61), I know what I want to teach but how can I know what they are going to learn?*

You've probably had similar epiphanies to Ralph's, or perhaps smaller moments of inquiry, where something trivial took on a greater sense of meaning. Perhaps it involved watching a bag of crisps expand on a mountain climb or marvelling at the enormous size of an oak tree, knowing that most of its mass came from the air around it and not the soil beneath it. What unites all of these experiences are moments of spontaneous inquiry, made possible by understanding a core set of scientific ideas.

▨ The big ideas of science

So, what are these scientific ideas that students should be learning about that have the potential to transform the bland everyday into moments of meaning?

Back in October 2009, a small group of scientists, engineers and science educators met on the shores of Loch Lomond in Scotland to explore this very question. At the heart of their undertaking was to identify a set of ideas that could frame the aims of science education. Each of these big ideas was a statement that represents some major scientific understanding that students should acquire by the time they leave compulsory education. An example of such an idea is that *'organisms are organised on a cellular basis'* (Harlen, 2010). By framing the aims of science education around big ideas, it was hoped that big ideas would overcome the problem of school science being perceived as a series of disconnected facts that just needed to be learnt (Harlen, 2010). In this way, each big idea acts as a 'peg' upon which students hang related knowledge and over time will see how small ideas, such as mitosis, protein synthesis, cell structure and cancer, are related through the bigger idea that *organisms are organised on a cellular basis*. Not only does building the curriculum around big ideas encourage students to make connections between concepts, it also encourages teachers to prioritise connecting and linking scientific concepts in their planning and teaching (Mitchell *et al.*, 2017).

In total, Harlen identified ten big ideas of science education (Harlen, 2010, 2015). Whilst these big ideas paved the way for re-thinking the aims of school science, they are not without their problems (Boxer, 2019).

Only one of the ten big ideas of science is specific to chemistry: *'all matter in the Universe is made of very small particles'*. This creates a number of problems if you want to use big ideas in school as in effect there is no big idea, there is only chemistry. By being so huge, the chemistry big idea is unable to stress the conceptually significant bits of the subject matter, such as entropy, substance and electrostatic forces (Gillespie, 1997). Another problem concerns some of the big ideas that relate to physics. Light and gravity are grouped together in Harlen's big ideas because they all act at a distance, however, this similarity feels quite superficial. Instead, it makes more sense from a conceptual point of view to understand light in relation to energy transfer, yet in Harlen's big ideas light and energy fall within separate big ideas. Whilst it is certainly true that *'the'* big ideas of science do not exist and so Harlen's big ideas are not wrong – it feels as if a more pragmatic set of ideas are needed that stress more explicitly the most conceptually important parts of science that we want students to acquire at school. It's time then to introduce you to what I am calling the powerful ideas of science, so named because they give 'power to' students, that form the focus of this book.

Powerful ideas of science

Slightly unnervingly, even for us scientists, I have selected 13 powerful ideas of science from across the domains of biology, chemistry and physics (see below). I have borrowed ideas from Harlen's original publication where possible and rephrased them in places to draw attention to conceptually important areas that should be stressed. For example, Harlen's original big idea that *'the diversity of organisms, living and extinct, is the result of evolution'* has become *'the diversity of organisms, living and extinct, is the result of evolution by natural selection'*. This change makes the distinction between evolution as a process – that is, a change in populations over time, and natural selection as the mechanism that brings about this change. This perhaps seems to be playing semantics, but as we shall see, meaning is everything when it comes to scientific ideas.

Where I felt I could improve upon Harlen's big ideas I did so. For example, magnetism and electricity have been brought together under one powerful idea, united by the concept of charge. Chemistry now comprises five powerful ideas instead of one, which I hope makes it easier to use these ideas to plan out a curriculum, ensuring that each idea is developed over time. Powerful ideas then act to *generate* the structure of the curriculum because they provide separate goals towards which we can sequence knowledge. Because the same powerful ideas can be used in different phases of education, it is easier to make sure that ideas introduced at one point in time are developed upon later. For example, teachers teaching about cells will build upon what students learnt about living and non-living objects in primary school, and to explain the mechanism of breathing, students will draw on what they know about gas pressure. In this way, the curriculum can be organised to facilitate progression within the same powerful idea *and* between different powerful ideas to give students the best possible chance of learning the scientific ideas in a meaningful way.

It's time then to introduce you to the powerful ideas of biology, chemistry and physics. For each powerful idea I have identified a question that students will be able to answer thanks to their new, scientific perspective. As students travel through their curriculum and learn more about these ideas, their answers to these questions will deepen, becoming increasingly sophisticated and meaningful. Of course, the true power of scientific ideas lies in the *many* questions that students can ask and answer because of them, so please don't see these questions as some aim of science education; they represent merely one star in the night sky.

*Powerful ideas of biology**

1. The cell is the basic structural and functional unit of life from which organisms emerge (informed by Nurse, 2003)
   Why don't bacteria have lungs?
2. Organisms reproduce by passing down their genetic information from one generation of organisms to another[†]
   Why are offspring similar but not identical to their parents?
3. Organisms compete with, or depend on, other organisms for the same basic materials and energy that cycle throughout ecosystems[†]
   Why are big fierce creatures rare?
4. The diversity of organisms, living and extinct, is the result of evolution by natural selection[†]
   How did giraffes get their long necks?

*Powerful ideas of chemistry**

5. Objects are made from materials, and materials are made from one or more substances built from atoms
   Are cars and organisms made from similar or different stuff?
6. When substances react, atoms are rearranged and new substances form but mass is always conserved
   What happens to the wax when you burn a candle?
7. Substances are held together by electrostatic forces of attraction
   Why does water stick together?
8. Chemical reactions only occur if they increase the disorder of the Universe
   Why do chemical reactions mess up our lives?
9. Quantities in chemistry are expressed at the macroscopic and sub-microscopic scales using grams, volumes and moles
   How much product can I make if I carry out this reaction?

*Powerful ideas of physics**

10. Changing the movement of an object requires a net force to be acting on it[‡]
    Why does coffee spill?

---

[*] *Numbers are to help organisation only and not intended to represent any sort of teaching order.*

[†] Adapted from Harlen (2010).

[‡] Taken from Harlen (2010).

11. The movement of charge forms electric current and causes magnetic fields

    Why don't birds resting on powerlines get electrocuted?

12. Every particle in our Universe attracts every other particle with a gravitational force

    How did planets form?

13. The total amount of energy in the Universe is always the same but can be transferred from one energy store to another during an event[‡]

    Why is it important to conserve energy resources if energy can't be destroyed?

Of course it is risky to break up science into 13 'tidy' categories like this, as it can deny the interrelatedness between concepts in different scientific ideas that clearly do exist. To understand why you should never make pineapple jelly, we need to think about proteases in pineapple, jelly being a mixture of sugar, protein and water and the attraction between the jelly to the Earth (and Earth to the jelly) that will ultimately cause it to collapse once the protein molecules in the gelatine get broken down by the proteases. However, the benefits of organising knowledge into separate ideas, I believe, outweighs the short-term costs because it allows students and teachers to specialise. Specialism not only helps teachers develop deep subject knowledge and manage workload (Sims, 2019), it also helps beginners to learn new ideas by reducing a vast discipline into a smaller number of related concepts. So, whilst interesting scientific advancements certainly do happen at subject boundaries (determining the structure of DNA involved a chemist and two molecular biologists), this is not a reason to destroy subject boundaries, it's a reason to keep them.

Scientific ideas as a story

Powerful ideas are a bit like a story in that we can't hope to understand them simply by reading their title. To understand a story, we need to know lots of knowledge about the characters, the setting and the plot and then we need to know the relationships between all of these parts. And, like a story, an understanding of a scientific idea requires that the knowledge is first defined, developed and connected in a logical order so that the relationship between the component parts is understood. If it's a good story it has the ability to

---

[‡] Taken from Harlen (2010).

transform the way we see ourselves and the world around us. Framing scientific ideas as stories has important pedagogical benefits (Landrum *et al.*, 2019). To see what this looks like, it's time for you to sit back, relax and read about one of the greatest stories ever told: the story of how you, a complex multicellular organism, emerged from a single cell some 10–100 years ago.

Powerful idea 1 *The cell is the basic structural and functional unit of life from which organisms emerge*

Like all good stories, let's start at the beginning. Objects can be living, non-living or dead. Rock is non-living, a chop is dead and an oak tree is alive. Living things do some marvellous things: they have sex, respire (transfer energy from food, usually by reacting it with oxygen) and respond to their environment. Organisms can do marvellous things because they have specialised organs (plants have sex and organs too; bacteria, though, just have sex). Organs can be further divided into tissues and cells; just like a house is made from bricks our bodies are made from cells – over 30 trillion of them. Cells, like bricks, can be specialised. The gangly neurone that descends over a metre from the base of your spine to your big toe contains exactly the same two-metre length of tightly coiled DNA (deoxyribonucleic acid) as the muscle cell in your leg, so how is it possible that these two cells look so different? We will return to this question in Chapter 3 after you have had some time to mull it over.

Take nothing for granted, you are a small miracle. Twenty-four hours after fertilisation, the fertilised egg cell (zygote) begins to divide in two by mitosis. One cell becomes two and two cells become four. Eventually, a ball of cells is formed which then differentiate to take on specific roles: some cells will go on to form the placenta (note this organ comes from mum and baby), whilst others will go on to form the embryo. Nine months later a baby (or babies) is born and cells will continue to divide until our death, and possibly for some minutes after.

Cell division is a complex and dynamic character in our story. On the one hand, it replaces worn out cells and enables babies to grow to adolescence, but when mitosis goes unchecked, cancer may result. Here your own cells now pose a risk to your very existence as uncontrolled mitotic cell divisions form invasive balls of cells that burrow deep into tissues, destroying their function. And when I say, 'your existence', we are of course talking about human *and* bacterial cells: our bodies are composed of broadly equal numbers of both (along with fungi and protists, too), although each defecation event may change the ratio somewhat.

Stories like these illustrate the joy that can be gained from understanding a scientific idea. But they also convey the power of a coherent science explanation, lesson or curriculum, where one idea builds from another. But rather than a story lasting a topic or term, powerful ideas transcend phases of education, with the 'key characters' such as cells or substances reappearing time after time, but in progressively more sophisticated and abstract ways enabling continuity (meeting the same characters) and progression (story getting progressively more complex). An understanding of many smaller ideas, such as living and non-living, the structure of animal and plant cells, eukaryotes and prokaryotes, and cell division, develops a scientific understanding that is bigger than the sum of its parts: the powerful idea has emerged and has enormous explanatory power, allowing us to explain:

- why hair falls out during chemotherapy – anti-cancer drugs target rapidly dividing cells that unfortunately includes hair follicle cells too;
- why bacteria don't need lungs – bacteria are single celled and have a sufficiently large surface area to volume ratio for diffusion alone to meet their requirements for gas exchange and
- why penicillin kills bacterial cells but not human cells – penicillin is an antibiotic that targets bacterial cell walls made from a substance called peptidoglycan. Human cells do not have cell walls and so are relatively unaffected by the drug. Plants are unaffected too because their cell walls are made from different stuff – cellulose.

An important goal: connectedness

We've so far described progression in understanding by focusing on one powerful idea. Our ultimate goal though, as we saw with the pineapple jelly, is for students to make connections between all of the powerful ideas of science, both within and between the three sciences. This interconnectedness enables truly powerful ways of seeing the world, for example, allowing us to appreciate the remarkable fact that you, I and this book are all made from stardust. You can explore this type of thinking in the classroom by asking students to connect up different concepts from different scientific ideas (Figure 2.1).

Inside **stars, elements** such as carbon, nitrogen and oxygen formed over billions of years. When these stars ran out of fuel they formed **supernova,** expelling **elements** into space, which then formed new stars or planets as the force of **gravity**[§] attracted the dispersed matter together. Eventually, some

---

[§] In this book I refer to gravity as a force according to Newton's law of universal gravitation. This allows us to describe and predict motion. If you want to explain why gravity acts, you will need to draw on Einstein's theory of general relativity that describes gravity not as a force, but as a consequence of the curvature of space-time.

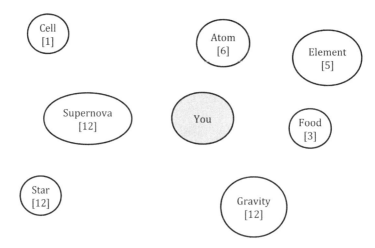

Figure 2.1 How are you connected to stars?

Note: Connecting concepts from different powerful ideas can explore students' understanding. Numbers in brackets correspond to the different powerful ideas described above.

4.5 billion years ago, some of these **elements** and the **atoms** they were made from formed Earth, ending up in your **cells** through the **food** you eat. So, we are just guardians for our atoms and who you are today is a different person from who you were as a baby – your atoms are not the same!

Summary

Scientific ideas are powerful ideas because they give students specialised ways of thinking about the world that they otherwise would not acquire through their everyday lives. These ideas are equally relevant to students who want to go on and work in science as they are for those who want to live with an appreciation of science. In this chapter, I have introduced you to 13 scientific ideas that stress the most conceptually important aspects of science for students and teachers at school. Acquiring an understanding of these ideas forms a major goal of science education and is reliant on students acquiring and connecting together lots of related knowledge bounded in concepts that are introduced and built upon in a logical way. But, as we will see in Chapter 3, an education in science is more than just understanding the scientific ideas. It is also about understanding how these ideas came to be in the first place because science is not just a body of knowledge, it's also a way of doing things that has its own rules and rituals that we want students to become familiar with.

Bibliography

Boxer, A. 2019. *What's the big idea?* Available at: https://achemicalorthodoxy .wordpress.com/2019/05/15/whats-the-big-idea/. [Accessed: 15 December 2019.]

Gillespie, R. J. 1997. The great ideas of chemistry. *Journal of Chemical Education*, 74(7), pp. 862–865.

Harlen, W., ed. 2010. *Principles and big ideas of science education*. Hatfield, UK: Association for Science Education.

Harlen, W., ed. 2015. *Working with big ideas of science education*. Trieste: IAP.

Landrum, R. E., Brakke, K. and McCarthy, M. A. 2019. The pedagogical power of storytelling. *Scholarship of Teaching and Learning in Psychology*, 5(3), 247–253.

Levinson, R. 2018. I know what I want to teach but how can I know what they are going to learn? Creative science teaching: An uncertain, emancipatory and perturbing endeavour, in *Critical issues and bold visions for science education*, Bryan, L. A. and Tobin, K. (eds.), 59–74. Leiden, The Netherlands: Brill Sense.

Mitchell, I., Keast, S., Panizzon, D. and Mitchell, J. 2017. Using 'big ideas' to enhance teaching and student learning. *Teachers and Teaching*, 23(5), pp. 596–610.

Nurse, P. 2003. The great ideas of biology. *Clinical Medicine*, 3(6), pp. 560–568.

Sims, S. 2019. *Increasing the quantity and quality of science teachers in schools: Eight evidence-based principles*. London: Gatsby Foundation. Available from: https://www.gatsby.org.uk/uploads/education /increasingscienceteachers-web.pdf. [Accessed: 15 December 2019.]

CHAPTER 3
The nature of science and the rules of the game

▇ Powerful idea focus

- *Organisms reproduce by passing down their genetic information from one generation of organisms to another.*

Less than 300 years ago people believed that inside every human sperm cell lay a homunculus, a miniature human being (Figure 3.1). Thanks to the work of many scientists, we now know that human sperm contain not a homunculus, but a set of 23 chromosomes. Our own genome comprises 46 chromosomes, arranged in 23 pairs – with one chromosome of each pair coming from a different parent. Along the length of these chromosomes sit genes that code for specific proteins that determine, to some extent, our characteristics. The nucleus then is analogous to a library, the chromosomes inside the nucleus are the cookbooks and the genes along the chromosomes are the recipes that code for our traits that we get from our parents. Current estimates suggest humans have somewhere between 19,000 and 20,000 recipes (or genes), although the precise figure seems to be eluding scientists, much to their frustration (Willyard, 2018). It seems organism complexity is not just about gene number, but concerns when these genes are switched on or off and in which combination.

Together, these ideas comprise the powerful idea of this chapter, that is, *organisms reproduce by passing down their genetic information from one generation of organisms to another*. But how do we know all of this? How do we know that chromosomes are inherited and that mini versions of our adult selves are not trapped inside our sex cells? I hope you're not just taking my word for it.

Figure 3.1 A homunculus inside a human sperm cell

Source: N. Hartsoeker, *Essay de dioptrique* available at https://wellcomecollection.org/works
/wsxjcdqe. Credit: Wellcome Collection. Attribution 4.0 International (CC BY 4.0).

The father of genetics

Gregor Mendel, botanist, monk and teacher, is the man responsible for much
of our modern-day understanding of inheritance. Working in the nineteenth
century, he spent many hours in the Augustinian Abbey gardens, in what
is now the Czech Republic, transferring pollen between thousands of pea
plants (*Pisum sativum*) using nothing but a paint brush. Other scientists had
done breeding experiments before, but Mendel's contribution stemmed
from the extensive and fastidious records he kept about the appearance and
number of offspring produced as he carried out his crosses.

When Mendel crossed-fertilised plants with wrinkled seeds with plants
with smooth seeds, he got smooth-seeded offspring (Miko, 2008). This
was strange, considering at that time inheritance was seen as a process of
blending: where were the semi-wrinkly seeds? The story got interesting
when Mendel observed the reappearance of wrinkled-seeded plants when
he cross-fertilised the smooth-seeded offspring from his first breeding experiment. Somehow the factor causing seeds to be wrinkled was present in the
parents, even though it wasn't seen in their appearance.

Mendel had found out something important. Factors, which we now call genes, come in different forms (alleles) that can be either dominant or recessive. You only need one dominant allele for the characteristic to be seen in the phenotype (appearance), whereas an organism needs two copies of a recessive allele for this trait to be visible. This means that recessive alleles can be maintained across generations even when they are not seen in the phenotype because they are being masked by the dominant allele. This is why redheaded children can be born from parents that have no red hair – parents simply need to be carriers for the recessive redhead allele that can then be passed to their children (there are other genes involved in this example, too). In the case of Mendel's peas, the allele for smooth seeds is dominant over the allele for wrinkled seeds. Inheritance then is not a process of blending, but rather involves the maintenance and shuffling of alleles between generations. If students are going to make sense of inheritance then they need to understand how alleles and genes are related – the multiple-choice question in Figure 3.2 is one way to explore their thinking.

Despite Mendel showing that inheritance was not a process of blending, the question remained as to how this shuffling process took place. Whilst scientists knew that chromosomes were the store of hereditary information, microscopes at this time struggled to provide clear images of chromosome movement during the formation of sperm and egg cells. This made it difficult to understand the mechanisms of inheritance that led to Mendel's results. That was until Walter Sutton (1877–1916), originally a farm boy from Kansas, made the fortunate discovery that chromosomes in the testes of lubber grasshoppers (*Brachystola magna*) were especially large. By studying meiosis

The diagram shows a pair of homologous chromosomes from a pea plant.

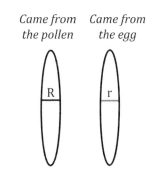

Select the *best* answer below.

The chromosomes:

a) are identical

b) contain different genes

c) contain different genes and different alleles

d) contain the same genes but different alleles

Figure 3.2 A multiple-choice question to explore the important difference between alleles and genes; the answer is (d) as one gene and two possible alleles (R and r) are shown

in this organism (cell division that produces gametes) Sutton was able to see the separation of the maternal and paternal chromosomes during cell division, leading to daughter cells inheriting only one of each chromosome pair. Sutton had provided a mechanism for the Mendelian laws of heredity:

> I may finally call attention to the probability that the association of paternal and maternal chromosomes in pairs and their subsequent separation during the reducing division ... may constitute the physical basis of the Mendelian law of heredity.
> *Sutton, 1902, p. 39*

The nature of scientific knowledge: laws and theories

The story of how the chromosome theory of inheritance came about illustrates some important principles in the way scientific knowledge is generated. Science is foremost an empirical endeavour where knowledge is generated by scientists observing and manipulating the natural world using a diversity of methods (Lederman *et al.*, 2014). In Mendel's case this was cross-fertilising pea plants and carefully observing and counting the progeny that led to the Mendelian laws of heredity. Laws like this are simply universal relationships or generalisations that describe how the natural world behaves under certain conditions. A law does not attempt to provide an explanation, it simply describes. In the case of Mendel, he had no understanding of the mechanism behind his observed ratios. It wasn't until Walter Sutton and others carried out their research (e.g., Theodor Boveri working with sea urchin eggs) that Mendel's observations could be explained, leading to the chromosome theory of inheritance.

Theories, then, are not stronger or weaker forms of a law, they are something different. Theories are inferred explanations of some aspect of the natural world that are supported by substantial empirical evidence and are accepted (National Science Teaching Association [NSTA], 2000) – they are not just a good guess in the way that we may have a theory as to why grandma fell over on the ice.

Like all scientific knowledge, laws and theories are only temporary and are revised in light of new evidence, which may support or refute existing hypotheses. So, whilst we tend to think of scientific ideas as right, for the most part they are only partially right and explain only some aspects of reality. Very few human traits are inherited in ways that Mendel's law predicts because few human traits are controlled by single genes. That's why a search for the gay gene or the gene for intelligence are probably intellectual dead ends. Complex traits like these are caused by multiple genes, interacting in complex ways with the environment. Often this revision in our scientific understanding is accompanied by the development of technology. Today, for example, rapid sequencing technology can sequence an entire human genome in less than a day (it took the human genome project 13 years, and that was despite the programme coming in two years ahead of schedule and under budget).

Theories are not the work of a single scientist either, but emerge from a combination of competition and collaboration between many scientists. Walter Sutton built upon the findings of Gregor Mendel and others have built on these findings over time. But it is because scientific discoveries are human endeavours that they are also subjective. What scientists do is affected by their beliefs, values and theoretical commitments that influence what gets measured. For example, Ronald Fisher, an initial sceptic of Mendel's results, only studied statistics and genetics because he was an ardent supporter of improving the human race through selective breeding – that is, he was a supporter of eugenics, as were many scientists at the time (Norton, 1978). That, of course, is not to say that Fisher ignored empirical evidence, just that what he chose to investigate was heavily influenced by social factors of the time.

Together, these ideas illustrate what is called the nature of science that describes the characteristics of scientific knowledge and how it is generated. These ideas are quite different from how scientists' work tends to be described in the classroom, where a single scientific method can quickly gain cult status. Not only does this oversimplification prevent students from appreciating the limits of scientific knowledge, it has the unintended consequence of portraying science as an inert discipline, devoid of human emotion and creativity that can put many students off from studying this subject beyond school.

The nature of scientific knowledge in the classroom

So, what does this look like in the classroom? A simple exploration into the nature of scientific knowledge could involve nothing more than a tin of beans with the label removed and a good question: *What is inside the tin?* Students will quickly see how their initial ideas are influenced by previous encounters with tins. There is then the opportunity to explore some of their ideas and build upon them. And no, scientists can't open the tin, only 'God' can. Instead, scientists must rely on making inferences. Possible ideas might include finding relationships between the unknown tin and some reference tins whose contents we do know (correlation), examining the codes and dates printed on the exterior (observation) or bombarding the tin with various X-rays and measuring how much of this is absorbed (experimentation). This initial, and yes, simplistic introduction to the nature of scientific knowledge can be built upon during subsequent years by honestly sharing how scientific ideas came to be. This moves thinking beyond seeing science as using a single scientific method that is free of subjectivity and error.

Scientific inquiry

Some people argue that young people only need to learn science at school and so we should stick to teaching the scientific ideas, they don't need to 'do science' (Kirschner, 2009). And when I say 'do science' I am not simply referring to practical work – I mean carrying out an investigation where students answer a question that matters to them, and where there is some choice in how this question is answered.

Criticism of scientific inquiries like this are justified for two main reasons. The first is that we can't expect students to generate knowledge new to science as only experts can do this. Whilst this is probably true for most, although it has happened (see, for example, work done by primary children looking at the vision of bumble bees [Blackawton *et al.*, 2010] or the work of a high school student in Tanzania, who identified that hot water freezes faster than cold water [Mpemba and Osborne, 1969]), very few scientists generate knowledge new to science either, instead spending much of their time measuring and recording data that agrees with what is already known. The second reason argues that learning scientific ideas whilst simultaneously using complicated and distracting equipment makes learning harder, as it overloads the limited processing capacity of our brains (Kirschner *et al.*, 2006). Whilst this is certainly true, using equipment such as Bunsen burners and filter funnels certainly complicates things, to some extent this criticism misses the point. Science is more than a body of knowledge about cells, atoms, forces and folding filter paper, it is also a discipline that uses this knowledge to inquire with and so students need the opportunity to experience both.

The primary purpose of scientific inquiry then is not to develop an understanding of concepts such as atoms or cells necessarily, although it could be, it is an education into the practice of science itself that allows students to experience and become acquainted with the 'rules of the game', often referred to as norms. This doesn't need to involve the generation of knowledge new to science but it should involve students generating knowledge that is new to them. Scientific inquiry then is foremost not a pedagogy but an opportunity for students to experience the pleasure of working like a scientist in a way that connects them to the discipline of science.

So, just as it would be absurd to study art but never allow students to paint a picture of their choice, it would be equally absurd to study science without having the opportunity to experience what it feels like to carry out a scientific inquiry. The trick, of course, is to plan the curriculum so that students have had the time to develop sufficient prior knowledge and skills to inquire with.

Scientific inquiry in the classroom

Many of the skills used in scientific inquiry, such manipulating apparatus, drawing tables and making observations, are probably best introduced in isolation, away from the distractions of Bunsen burners, breakable glassware and overly complicated concepts. This doesn't mean students need to know everything before they begin, though – some learning will take place during the inquiry itself – just that there needs to be sufficient scaffolding (Hmelo-Silver *et al.*, 2007) and the opportunity and time to acquire any knowledge students don't have by providing access to:

- textbooks
- selected resources, e.g., YouTube videos
- lab protocols
- peers
- teacher feedback

Students can either generate their own inquiry questions or they can be given a question where there is an element of choice in how this is answered. Possible ways to stimulate inquiry questions can involve bringing in artefacts such as objects, newspaper reports, graphs, tables of data or pictures. Students' questions can be explored on the board together as a class before refining these further. You can create the feeling of choice in how the inquiry is done by providing a range of different apparatus, methods or reagents to use – perhaps including more than they will need. The focus, at least in initial inquiries, is on the illusion of choice. Selecting a 'safe' practical context for the inquiry can help here as it gives students freedom to make mistakes, without risk of significant harm.

Table 3.1 provides some ideas of inquiries that you could try, progressing from closed to more open ones. You can see that they don't always need to take up a lot of time, but they should provide results that can be explained using scientific ideas that students understand.

Table 3.1 Some suggested scientific inquiries

Scaffolding	Question	Stimulus provided by the teacher	Background	Concepts in curriculum
High	How could you store this piece of string so it occupies the smallest amount of space?	Give each pair of students a 2 m piece of string and access to scissors. The string represents the total length of double-stranded DNA in a human cell.	This short activity (5 mins max) provides an introduction into the challenges of storing genetic information inside a tiny cell. It helps students appreciate the need for DNA to be coiled and packaged into chromosomes. You can then use this model in later lessons when exploring the relationship of genes to chromosomes.	Chromosomes, genes, mitosis, meiosis, nucleus.
Medium	Students to decide on the question when given a stimulus.	A range of fruits, detergents, proteases and water baths set at different temperatures.	Students have already been shown the method to extract DNA from kiwi fruits and have successfully completed this practical activity. They then use this prior learning in the inquiry.	DNA, enzymes, nature of science.
Low	Students to decide on the question when given a stimulus.	A selection of leaves from the same and different plant species, or students could collect their own leaves from the local park.	Students to decide—examples: analysis of pigments using chromatography, strength, stomatal density, water content, surface area, arrangement.	Variation, inheritance, photosynthesis, water transport, adaptation.

■ Applications of science and their implications for society

The products of scientific inquiry though don't just reside inside tables of results, test tubes and textbooks, they also have a number of applications that themselves have implications in the 'real world'. To illustrate what I mean, let's return to the mid-1990s in America, where an important socio-scientific issue involving a gene, some patients and a diagnostic company made it to the Supreme Court of the United States.

In healthy human cells, the genes BRCA1 and BRCA2 play important roles coding for proteins that repair damaged DNA. However, a mutation in either of these genes creates alleles that significantly increase the risk of developing breast and ovarian cancer. Therefore, the potential to detect mutated versions of BRCA1 and BRCA2 was an important goal for diagnostic companies in the mid-1990s for both medical and financial reasons. Whilst a number of organisations set out to locate and clone these genes, it was Myriad Genetics, an American diagnostic company based in Salt Lake City, who got there first and subsequently filed for patents (Cartwright-Smith, 2014).

With their patents of BRCA1 and BRCA2 approved, Myriad Genetics moved to stop other commercial organisations from offering similar tests. However, the legitimacy of these patents was soon challenged, with the Federal District Court declaring that the patent was invalid, citing that a gene is a product of nature and so cannot be patented. This judgement was not to last long and was subsequently overturned on appeal after another ruling on a similar case. At this time Myriad Genetics was charging somewhere between $3,000 and $4,000 per test (Cartwright-Smith, 2014). A new appeal, led by the Association for Molecular Pathology, was taken to the US Supreme Court in 2013, which ruled 9 to 0 that isolated genes could not be patented but synthetic DNA could be. This ruling allowed other companies to market the BRCA1 and BRCA2 tests and the cost of the test fell dramatically.

This case illustrates how the applications of scientific knowledge, in this case the diagnostic test for specific forms of cancer, can have profound effects for society, creating winners and losers. Here, the winners were those patients who could afford the test and the losers those who couldn't. Interestingly, it wasn't the application itself that was problematic, but the implications that stemmed from it. You find similar examples of winners and losers in pretty much every application of science (Levinson, 2018). Chlorofluorocarbons (CFCs) were introduced as safer, non-flammable alternatives to existing refrigerants, but they soon turned out not to be such a good idea, causing depletion of the ozone layer, which in turn caused an increase in incidences of skin cancer.

To appreciate the implications of scientific knowledge then, we can't just confine school science to thinking about the closed system that operates inside test tubes, we also need to consider the political, social and economic consequences that stem from the scientific ideas; in the case of Myriad Genetics, this was simply the implication of breaking covalent bonds needed to remove the BRCA1/2 genes from the genome.

Applications of science and their implications for society in the classroom

We can help students to see how the applications of science have important consequences for society by drawing on real-life examples. This could involve students making judgements on issues by weighing up the advantages and disadvantages of specific technologies by considering the implications of the scientific ideas. For example, should you be able to patent genes, clone animals or vape in public (Figure 3.3)? A writing frame with a scenario can help focus thinking but as with inquiry, students need to have, or be motivated to find out, sufficient scientific ideas to think from. It would be impossible to debate whether the US Supreme Court made a fair and valid judgement without an understanding of covalent bonds and gene structure. You can collate views from tasks like these by asking students to place a counter in one of two pots as they leave the classroom and then share the results at the start of the next lesson. This doesn't need to take ages, rather it

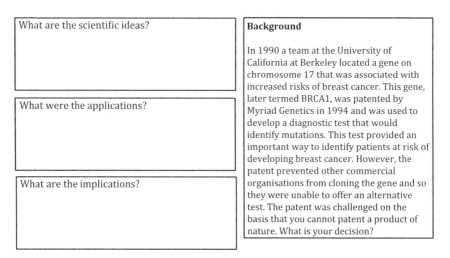

Figure 3.3 A simple writing frame to help focus student thinking on how the applications of science can have implications beyond the science classroom

helps to make the point that scientific ideas matter beyond the classroom, as do students' opinions.

You could focus thinking further in other lessons by looking at who the winners and losers are for different socio-scientific issues such as genetic modification, nuclear fuel and the extraction of metals. The trick then is to find the dilemma that is best understood by following through what the scientific ideas mean for different 'characters' within the story. It is through learning about and understanding these 'interlocking narratives' (Levinson, 2009) that students will form a more genuine understanding of the scientific concepts and the implications that stem from them.

Summary

A science education is not restricted to learning scientific concepts, it's also about having the opportunity to consider how and why these concepts came to be. By understanding the stories behind the powerful ideas of science, students can see the limits and potential of scientific knowledge and have a more genuine understanding of science as a tentative, subjective and empirical human endeavour. By taking part in scientific inquiries, students experience first-hand this tentative and subjective nature but they also get to connect with the 'practice of science' by exploring questions that matter to them in ways that foster positive feelings of autonomy and competence. Finally, students need the opportunity to see how science doesn't just operate inside laboratories but has profound implications for society, both good and bad. Together the ideas discussed in this chapter comprise the nature of science and explain how and why scientific knowledge is generated in the first place. As we will see in Chapter 4, learning these scientific ideas, even with careful instruction, is not always that easy.

Bibliography

Blackawton, P. S., *et al*. 2010. Blackawton bees. *Biology Letters*, 7(2), pp. 168–172.

Cartwright-Smith, L. 2014. Patenting genes: What does Association for Molecular Pathology v. Myriad Genetics mean for genetic testing and research? *Public Health Reports*, 129(3), pp. 289–292.

Hmelo-Silver, C. E., Duncan, R. G. and Chinn, C. A. 2007. Scaffolding and achievement in problem-based and inquiry learning: A response to Kirschner, Sweller, and Clark (2006). *Educational Psychologist*, 42(2), pp. 99–107.

Kirschner, P. A. 2009. Epistemology or pedagogy, that is the question, in *Constructivist instruction: Success or failure?* Tobias, S. and Duffy, T. M. (eds.), 144–157. New York, NY: Routledge.

Kirschner, P. A., Sweller, J. and Clark, R. E. 2006. Why minimal guidance during instruction does not work: An analysis of the failure of constructivist, discovery, problem-based, experiential, and inquiry-based teaching. *Educational Psychologist*, 41(2), pp. 75–86.

Lederman, N. G., Antink, A. and Bartos, S. 2014. Nature of science, scientific inquiry, and socio-scientific issues arising from genetics: A pathway to developing a scientifically literate citizenry. *Science and Education*, 23(2), pp. 285–302.

Levinson, R. 2009. The manufacture of aluminium and the rubbish-pickers of Rio: Building interlocking narratives. *School Science Review*, 90(333), pp. 119–124.

Levinson, R. 2018. Realising the school science curriculum. *The Curriculum Journal*, 29(4), pp. 522–537.

Miko, I. 2008. Gregor Mendel and the principles of inheritance. *Nature Education*, 1(1):134. Available at: https://www.nature.com/scitable/topicpage/gregor-mendel-and-the-principles-of-inheritance-593/. [Accessed: 15 December 2019.]

Mpemba, E. B. and Osborne, D. G. 1969. Cool? *Physics Education*, 4, pp. 172–175.

National Science Teaching Association (NSTA). 2000. *NSTA position statement. The nature of science.* Available at: https://www.nsta.org/about/positions/natureofscience.aspx. [Accessed: 15 December 2019.]

Norton, B. 1978. A 'fashionable fallacy' defended. *New Scientist*, 78(1100), pp. 223–225.

Sutton, W. S. 1902. On the morphology of the chromosome group in Brachystola magna. *Biological Bulletin*, 4, pp. 24–39.

Willyard, C. 2018. *New human gene tally reignites debate.* Available at: https://www.nature.com/articles/d41586-018-05462-w. [Accessed: 15 December 2019.]

Section 2

The science of learning science

CHAPTER 4

Why learning science is hard but wonderful

Powerful idea focus

- *Changing the movement of an object requires a net force to be acting on it.*

Imagine, for a second, that you are flying 30,000 feet above sea level at a speed of 500 mph. The seat belt sign has just been switched off and, as this is a long-haul flight, you get up from your seat to have a stretch. As you're standing in the aisle you take a jump, accelerating straight up into the air. Where will you land? Intuitive thinking suggests that you will end up closer to the back of the plane, but should you trust your intuition? The answer is probably not, and as we explore the powerful idea of this chapter, that is, *changing the movement of an object requires a net force to be acting on it*, we will explore the reasons why.

Not all knowledge is equally easy to learn

To understand why we shouldn't always trust our intuition, we need to think about the types of knowledge that we learn about during our lifetime. Broadly speaking, we can think of knowledge as falling into one of two main types: biologically primary knowledge and biologically secondary knowledge (Geary, 1995). Biologically primary knowledge is associated with behaviours such as learning to speak, imitating gestures from our parents and recognising faces. Over millions of years, we have evolved sophisticated mechanisms to acquire this type of knowledge, because there was an evolutionary cost in not learning it; either we lost our parents, starved, failed to

secure a mate or got eaten by something. Learning associated with biologically primary knowledge takes place automatically, with minimal effort and
can be quite fun (Kirschner *et al.*, 2018)! Play, for example, is a behaviour that
takes place in a variety of species and has an important role in developing
adult-like abilities to acquire biologically primary knowledge.

However, there is no selective advantage in knowing that the equation to
calculate force involves multiplying the mass of an object by its acceleration.
Or, thinking about this in a different way, not knowing the equation for force
doesn't affect your likelihood of survival and/or reproduction. In fact, there
may well be an inverse correlation here! This type of knowledge, referred to
as biologically secondary knowledge, can be quite tricky to learn and doesn't
have the same immediate enjoyable associations that learning biologically
primary knowledge does. Take, for example, learning to read. Learning to
read is a form of biologically secondary knowledge and it's hard, requiring
explicit instruction where words must be broken down into phonemes
(sounds) and graphemes (letter groups). Despite huge financial investments,
UNESCO (2017) reports that over 617 million children worldwide still cannot
read a basic sentence, showing just how difficult it can be to learn biologically secondary knowledge. Formal school science requires students to learn
such biologically secondary knowledge and to make matters worse, students
are not learning these scientific ideas in school from scratch.

Naive ideas: we are condemned to learn

As young children navigate their place in the world, they interact with many
things through their senses: banging onto tables, rolling marbles down a
slope and sucking on pebbles. The world is a chaotic place that needs to be
ordered. To create order, or make sense of this world, children use quite a bit
of reasoning based on cause and effect which develops early on in childhood
(Pinker, 1999). Children will push a toy car and it will move. When they
stop pushing the car, it stops. Over time these reasoning patterns allow children to build, or construct in their minds, personal understandings that
psychologists call naive ideas; sometimes these ideas are correct but often
they are wrong. Unfortunately for science teachers, many of these wrong
ideas still work! Cars do stop moving when a force is no longer applied and
heavy objects do fall faster than light ones.

Ideas that differ from the scientific way of thinking have been called a
variety of things, from naive conceptions to alternative conceptions to
misconceptions or private theories. How you name these ideas ultimately
depends on your perspective. They are misconceptions in that these ideas

are incompatible with the scientific way of thinking, but at the same time they are not always incompatible with day-to-day lives. On Earth, rest does seem to be the natural state of objects as balls do stop moving because of friction (although friction is also absolutely necessary to get things moving in the first place!). In this book, we will use the term 'misconception' to refer to these unscientific ideas that differ from the scientific view, simply because many teachers recognise and use this word in schools. However, it would be wrong to take from this that misconceptions are somehow based on faulty logic in a way that mistakes in maths or reading might be. Indeed, misconceptions are often far more logical than their scientific counterparts.

A common example of a misconception in physics is that a force is needed to keep an object moving, often called the impetus theory. Most children (and many adults) believe that objects will only move when a force is applied, after all, this is what we experience (Driver *et al.*, 1994). Balls appear to move through the air because a force is transferred from your hand to the ball, almost as if you've attached an invisible engine. But this common sense explanation does not agree with the scientific idea. Once the ball has left your hand there is no force pushing it, so why then does it continue to move? Perhaps a better question is why does it not stop? We can explore this reasoning using Galileo's thought experiment involving a marble moving along a frictionless surface (Figure 4.1). In the first picture you can see that a ball released at the top of a V-shaped slope will rise to an equivalent height that it started from – energy is conserved. But where will the ball end up in the final picture? The answer is, assuming no frictional force, it will go on forever. Forces are only needed to change the direction or motion of an object, they are not required to keep an object moving.

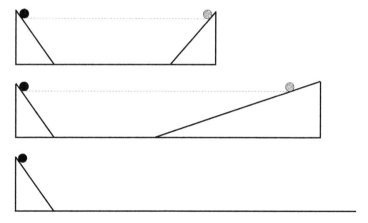

Figure 4.1 Galileo's thought experiment involving rolling balls down frictionless hills

Children's naive ideas are hardly surprising when you consider it took scientists hundreds of years to figure out these scientific ideas. Newton's first law of motion, that is, '*an object at rest stays at rest, and an object in motion stays in motion with the same speed and in the same direction unless acted upon by an unbalanced force*' (The Physics Classroom, n.d.) wasn't published until 1687. Students aren't going to discover Newton's first law in a world with friction unless given some help, and it's not just children who hold misconceptions, either.

Sketch, for a moment, what happens to a ball travelling at 50 mph that rolls off the edge of a frictionless cliff and label the significant forces acting.

If you've drawn anything other than a parabolic trajectory, with a down-ward force, you're wrong, but in good company (McCloskey, 1983). Once the ball leaves the cliff it will continue to travel forwards at a constant velocity, but importantly, no force is acting in this direction (Figure 4.2). The down-ward velocity of the ball is, however, increasing due to the force of gravity on the mass of the ball, which on Earth accelerates all objects, regardless of their mass, at 9.8 m/s². We will look further at gravity in Chapter 11 to see how this is possible.

Let's return then to our thought experiment at the start of this chapter. The passenger who jumps up in the aisle of the plane will land exactly where they started from because no force was applied in a horizontal direction. Or explained in a different way, a passenger travelling at 500 mph will continue to travel at 500 mph unless a force is applied, irrespective of whether they leave the floor or not.

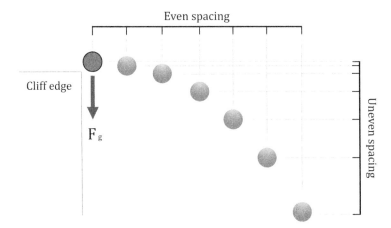

Figure 4.2 Consider both the horizontal and vertical components of motion when a ball rolls off a cliff

Misconceptions in biology

Similar naive reasoning happens in biology too. Four-year olds understand that living things will re-grow, whereas other objects, such as pottery, need to be mended by humans (Backscheider *et al.*, 1993). By the age of six, children will begin to classify objects based on their function and children will expect living things to behave differently from non-living things.

Whilst naive ideas in biology seem less problematic than those of physics, in that these ideas are not so different from the scientific ones, naive biology still poses a few problems. When children have incomplete information, they may personify organisms by giving them human attributes. So, if a person taking care of a grasshopper suddenly dies, children suggest that the grasshopper 'may feel unhappy' (Hatano and Inagaki, 1994); perhaps they do. Children may also take quite a teleological perspective, that is, they tend to explain natural phenomena by reference to a purpose, often for human benefit. Bacteria mutate to become drug resistant and earthworms tunnel to aerate the soil (Kelemen and Rosset, 2009). It would be wrong, however, to dismiss teleological thinking as always being an obstacle to science. Darwin was a teleologist right through *On the Origin of Species* and Newton ascribed a superior intelligence to planetary motion. Indeed, teleological thinking can be helpful for students, at least in the initial stages of learning scientific ideas, where abstract ideas can be made to appear more tangible and relatable. The goal though, with any approximation like this, is for students to be aware of its limitations.

Misconceptions are not just errors

Many of these misconceptions make school science quite a confusing place. What was once a useful conception that explained why marbles stop rolling along a table is now an error marked wrong on a force diagram. Science lessons, then, have to involve learning, refining and unlearning ideas (Shtulman and Harrington, 2016).

Misconceptions, though, are not just errors. Students can construct quite elaborate and coherent explanations for their unscientific ideas, drawing on aspects of correct science and integrating these with their naive ideas to form a personal explanation or theory. Take, for example, the idea of floating and sinking, a phenomenon that many children (but not all, as some only have showers) will have experienced at bath time. Children can explain floating by referring to concepts of weight and upthrust (the upward push of water acting on a floating object), but wrongly believe that two objects of the same weight will both float because they will produce the same upthrust. But take two identically sized pieces of plasticine, roll one into a ball and flatten the other into a boat shape and students will see for themselves that weight alone is not sufficient to explain why objects float. For an object to float, the downward weight force on an object must be balanced by the upward push from the water (upthrust). When you flatten the plasticine into a boat shape, it pushes away more water, creating a larger upthrust that balances the downwards weight force. So, although the student's explanation is wrong, two objects with the same weight don't necessarily float, there is quite a sophisticated explanation that is rooted in many logical (but erroneous) ideas about forces. And to make matters tricky for science teachers, these ideas are often only subtly different from the scientific way of thinking.

Knowledge in pieces

So far we have described misconceptions as deeply held theories that we hang on to at all costs. Over the years there has been a debate rumbling on within the science education community as to whether students reason in science from theory-like structures, or instead *ad hoc* explanations that are assembled in that instance from fragmented knowledge. It's the difference then of having a stable belief about a wrong idea versus just getting it wrong in a particular situation when the answer is constructed. This second way of thinking about misconceptions is the 'knowledge in pieces' argument and was made by Andrea diSessa (1993). It considers that misconceptions emerge from scraps of knowledge called phenomenological primitives, or p-prims, that are used to explain many different, seemingly unrelated situations. Examples of p-prims include *closer means stronger* or *dying away* (Hammer, 1996). Students can use *closer means stronger* to incorrectly explain why summer is hotter than winter (Earth is not closer to the Sun during summer months), but the same p-prim correctly explains why you don't want to get too close to a fire. In reality, the truth probably lies somewhere in the middle, in that students use a mixture of theory-like and p-prim explanations to explain the world. The good news for teachers is whether misconceptions are deeply held or formed from knowledge in pieces, advice on how to resolve them is not that different.

Addressing misconceptions

To address misconceptions in the classroom, there are a number of strategies we can try. These include:

- anticipating misconceptions before teaching, for example, by consulting Rosalind Driver's work on misconceptions (Driver *et al.*, 1994),
- identifying existing ideas, for example, using Galileo's thought experiment,
- building from students' existing ideas, even if they are wrong,
- showing students why their wrong ideas don't work and so need modifying, for example, we could ask a student holding the impetus theory of motion to imagine a cup of coffee being held in a car that suddenly stops. If forces are needed to keep objects moving then coffee shouldn't spill, but it does,
- exploring similar ideas across a range of contexts – don't just explore Newton's first law by looking at moving objects on Earth, take an imaginary trip to space,
- providing lots of opportunities to practice using and applying the scientific ideas to a range of different situations, for example, ask students to draw force diagrams for a variety of similar situations and identify the forces acting on specific objects and
- setting problems that require students to apply this understanding, for example, to explain why seat belts save lives during a collision.

Research suggests that both the old (misconception) and new conception (scientific idea) are probably retained after instruction (Shtulman and Valcarcel, 2012), a bit like a dormant virus ready to 'jump out' at any time! Alzheimer's patients, who have difficulty inhibiting ideas, often support teleological explanations for natural phenomena in ways that young children do and even university biology professors are slower to classify plants as living than they are to classify animals as living (Goldberg and Thompson-Schill, 2009). It seems then that our focus should be on helping students to select the most useful conception when solving a problem (Ohlsson, 2009). Over time, as they develop expertise in the subject, they get better at suppressing the misconceptions and selecting the scientific ideas. Much of this expertise is developed through practicing and applying ideas as described earlier. This practice develops knowledge of the scientific concept and allows students to see the limitations of their previous ideas, it also inducts students into the norms of scientific discourse. In other words, as students learn about the scientific ideas, they also learn how scientists talk about things and how this is different to everyday ways of talking. As they learn to talk the language of science, students' explanations become more precise and resemble more closely the scientific way of thinking. We will explore further ways to address misconceptions throughout this book, but for now recognise that they are stubborn, often functional and are probably context-specific. And remember, a wrong answer is not always evidence of a misconception or p-prim, it may simply be a consequence of having nothing better to say.

Other reasons why learning science is hard

Learning science isn't just hard because of misconceptions, though (Johnstone, 1991). Let's take a look at some of the other reasons for why students can find science a tricky subject at school.

Scientific ideas are counterintuitive, going against everyday experiences

When a car is stationary you might assume that no forces are acting on it, but there are. In this case there is a push of the ground on the car and a pull of the Earth on the car. These forces are balanced and this is why the car does not move. A state of rest then is not that different from a state of constant motion in that both situations involve balanced forces.

Scientific ideas involve thinking at many different levels, all at the same time

To understand friction, we need to think about and move between different levels of thought (Johnstone, 2007): starting at the macroscopic and familiar (two objects in contact with each other) to symbolic/mathematical (force diagrams showing grip and slip forces) to the submicroscopic (electrostatic attractions between the two surfaces). In biology, this also involves thinking about the microscopic (cells) level too.

Experts are quite good at moving between these levels when thinking about science, but novices, that is, students with little prior knowledge, find this much more difficult and need some help by explicitly showing them where in the rectangle they are (Figure 4.3). For example, a circle drawn on the board could represent a particle, zero, a cell or an orange!

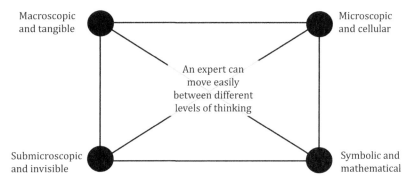

Figure 4.3 The four different levels that scientists use to think about the world

Source: Adapted from Johnstone, 2007.

Scientific ideas can't be seen directly and so must be inferred from something else

We can't see air particles or forces, instead we must make inferences about them from something else. For example, put your face near a fan and you will feel the wind hitting your head. Similarly push a football underwater and you will feel the effects of the upthrust force.

Scientific ideas are communicated using a whole new language that is different and distant from everyday words

The weight of an apple is not 102 grams, it's 1 newton. Learning science requires students to learn more words than they do in foreign language classes, but with the added difficulty that in science, many of the words, such as force or cell, have different meanings to those used in everyday life. Even the representations we use in science can have different meanings. Take the humble arrow: is it a force, a transfer of energy or a chemical change? It is the context which determines the meaning of the arrow, which is obvious for a science teacher but less so for a student who has just come from their Geography lesson, where arrows point North.

Scientific ideas are hierarchically arranged

We understand scientific concepts by knowing lots of things about them, and these concepts tend to be arranged hierarchically. To understand why crossing the Sahara in your stilettoes is a bad idea you need to know:

1. That a stiletto shoe has a long, thin, high heel.
2. That the Sahara is mostly sand.

3. That sand is a solid but can behave like a fluid.
4. That a person has a mass.
5. That there is a pull on the person and their shoe by the Earth.
6. This pull is called weight.
7. Weight is a force measured in newtons.
8. That weight spread over a smaller surface area produces a larger pressure.
9. Objects that produce a large pressure sink into the sand.
10. Pressure is force per unit area or $pressure = \dfrac{force}{area}$.

To understand each idea you really need to understand the one that came before. Whilst it's likely that students will move up and down this 'ladder' of progression as they struggle to make sense of the idea, if they get stuck on any one of these rungs then conceptual learning will stop. This makes identifying misconceptions and scaffolding learning especially important in science.

Learning science is wonderful

I hope the case has been made that learning science is hard, but why is it wonderful? Well, here is the paradox: learning science is wonderful because learning science is hard. Within every science lesson there is an opportunity to take students beyond their everyday experience and give them a radically different way of thinking about their world. To understand why a bullet fired from a gun and a bullet dropped from the same height will reach the ground at the same time is mind-blowing and wonderful. This counterintuitive idea only makes sense when you understand that the pull of the Earth on the bullet is the only significant force acting on both bullets in the air. The fired bullet is **not** being pushed along once it has been fired as our intuition would suggest! This makes learning science truly transformative because scientific ideas open up entirely new ways of thinking about the world for all students: objects don't move because you push them, objects move regardless!

Summary

Learning science is hard because scientific ideas are 'crazy ideas' that are abstracted away from our everyday experiences. Scientific ideas are counterintuitive, involve thinking at many levels, are described using a new language and require prior ideas to be securely understood before new learning can take place. This causes learners to construct quite complex but incorrect ideas about the world prior to and during instruction, meaning that novice learners both know less and *think differently* to experts. At the same time, it is because science offers a dramatically different way of thinking, which conflicts with much of what we already know, that learning science can be wonderful and truly transformative in ways that other subjects can't. However, we cannot take such an appreciation of science for granted. For many students, school science represents a body of fragmented facts that need to be learnt, that are neither useful nor interesting. Fragmented knowledge is not powerful and so we need to help students to acquire organised knowledge, where new ideas are incorporated under the meaning of pre-existing ones. The challenge then is how we take a curriculum, filled to the brim with counterintuitive and abstract concepts, to the classroom in a way that leads to meaningful learning *for all*. To be in with a chance, we are going to need to explore how students learn.

Bibliography

Backscheider, A. G., Shatz, M. and Gelman, S. A. 1993. Preschoolers' ability to distinguish living kinds as a function of regrowth. *Child Development*, 64(4), pp. 1242–1257.

diSessa, A. A. 1993. Toward an epistemology of physics. *Cognition and Instruction*, 10(2–3), pp. 105–225.

Driver, R., Squires, A., Rushworth, P. and Wood-Robinson, V. 1994. *Making sense of secondary science: Research into children's ideas*. London: Routledge.

Geary, D. C. 1995. Reflections of evolution and culture in children's cognition: Implications for mathematical development and instruction. *American Psychologist*, 50, pp. 24–37.

Goldberg, R. F. and Thompson-Schill, S. L. 2009. Developmental "roots" in mature biological knowledge. *Psychological Science*, 20(4), pp. 480–487.

Hammer, D. 1996. Misconceptions or p-prims: How may alternative perspectives of cognitive structure influence instructional perceptions and intentions. *The Journal of the Learning Sciences*, 5(2), pp. 97–127.

Hatano, G. and Inagaki, K. 1994. Young children's naive theory of biology. *Cognition*, 50(1–3), pp. 171–188.

Johnstone, A. H. 1991. Why is science difficult to learn? Things are seldom what they seem. *Journal of Computer Assisted Learning*, 7(2), pp. 75–83.

Johnstone, A. H. 2007. Science education: We know the answers, let's look at the problems. In *Proceedings of the 5th Greek Conference 'Science education and new technologies in education'*, 1, pp. 1–11.

Kelemen, D. and Rosset, E. 2009. The human function compunction: Teleological explanation in adults. *Cognition*, 111(1), pp. 138–143.

Kirschner, P. A., Sweller, J., Kirschner, F. and Zambrano, J. 2018. From cognitive load theory to collaborative cognitive load theory. *International Journal of Computer-Supported Collaborative Learning*, 13(2), pp. 213–233.

McCloskey, M. 1983. Naive theories of motion, in *Mental models,* D. Gentner and A. L. Stevens (eds.), 299–324. Hillsdale, NJ: Erlbaum.

Ohlsson, S. 2009. Resubsumption: A possible mechanism for conceptual change and belief revision. *Educational Psychologist*, 44(1), pp. 20–40.

Pinker, S. 1999. How the mind works. *Annals of the New York Academy of Sciences*, 882(1), pp. 119–127.

Shtulman, A. and Harrington, K. 2016. Tensions between science and intuition across the lifespan. *Topics in Cognitive Science*, 8(1), pp. 118–137.

Shtulman, A. and Valcarcel, J. 2012. Scientific knowledge suppresses but does not supplant earlier intuitions. *Cognition*, 124(2), pp. 209–215.

The Physics Classroom. n.d. *Newton's first law.* Available at: https://www .physicsclassroom.com/class/newtlaws/Lesson-1/Newton-s-First-Law. [Accessed: 15 December 2019.]

UNESCO. 2017. *More than one-half of children and adolescents are not learning worldwide.* Available at: http://uis.unesco.org/sites/default/files /documents/fs46-more-than-half-children-not-learning-en-2017.pdf. [Accessed: 15 December 2019.]

CHAPTER 5

How we learn
A cognitive science perspective

Unlike fats or carbohydrates, humans cannot store excess protein. Instead, amino acids, the building blocks of proteins, are broken down in the liver to form urea. This is then excreted from the body in urine in quite substantial amounts (25–40 g/day). Despite urea being the precursor to the rather unpleasant smell that lingers over urinals (urea is broken down into ammonia by bacteria), it is a rather significant substance (Figure 5.1). It was in 1828 that the German chemist Friedrich Wöhler first synthesised urea in the laboratory, putting an end to the idea that life was made from different stuff than non-living objects.

All materials then in the Universe, both living and non-living, are made from the same basic elements that make up all substances we see, taste and touch. A substance can be either an element or a compound (Figure 5.2) and is a fundamental concept in chemistry that is often overlooked (Johnson, 2011). This means many children see the world as being made from solids, liquids and gases, instead of substances such as carbon, nitrogen and carbon dioxide. Carbon, nitrogen and oxygen are all elements because they are made from only one type of atom and can be found on the periodic table. There are 118 known elements in the Universe and they can all be explored at: http://www.periodicvideos.com/.

Urea is a compound because it is made from different elements (carbon, hydrogen, nitrogen and oxygen) chemically bonded together. Urine is neither

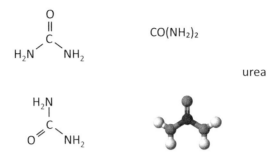

Figure 5.1 Just some of the different ways that chemists represent urea

a compound nor an element but is a mixture of different substances, including water, urea and sodium chloride, that together form a solution. Mixtures like urine can be separated out because they are made from substances which have different physical properties; in this case evaporation followed by condensation will give us water good enough to drink as all other substances, with their comparatively high boiling points, are left behind. Thanks to this principle of distillation, crewmembers on the International Space Station are able to get water from their waste.

Conceptual change

The process by which students build new ways of thinking from old ones is called conceptual change (Treagust and Duit, 2009). This change involves

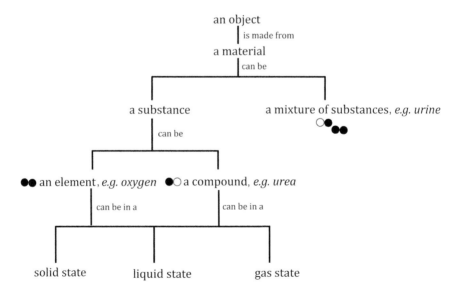

Figure 5.2 How chemists think about materials and substances

modifications to beliefs, theories and concepts, leading from a shift in novice understanding to an expert understanding as ideas are discussed, refined and accepted. So, whilst students may start off struggling to distinguish between an object and the materials it is made from, they will later see, with some help from their teacher, that objects, such as candles, are made from materials, such as wax, and that these materials are made from one or more substances built from atoms. Wax then is an example of a mixture because it contains lots of different substances such as $C_{25}H_{52}$ and $C_{30}H_{62}$ that are not chemically joined together.

These personal journeys towards understanding scientific ideas often parallel the development of scientific ideas throughout history – descent with modification as Darwin would say. Many students start off seeing atoms as tiny spheres, just as John Dalton (1803) did, but later will refine their model, seeing atoms more like solar systems than snooker balls, resembling the model proposed by Niels Bohr in 1913. Here electrons revolve around a heavy nucleus like planets orbit our Sun. Matter then, like our solar system, is mostly empty space. Inside the heavy nucleus sit positively charged protons and neutral neutrons which are responsible for an atom's mass. In the case of hydrogen, the nucleus contains no neutrons and only one proton, which, together with its explosive nature with oxygen, explains why it is an excellent choice for rocket fuel – it has an extremely low density. It is the number of protons inside the nucleus, and not the neutron number, that determine what the atom is. Carbon has six protons, whilst uranium, that we met in Chapter 1, has an impressive 92. But it is the outer region of atoms that get chemists most excited because chemical reactivity is influenced by the arrangement of electrons on the outermost shell.

Despite the apparent similarities in how scientific ideas evolve in learners' minds and throughout the history of science, learning science can be harder than doing science. Whilst scientists make sense of phenomena using their pre-existing scientific ideas, students learning science at school are stuck in a quandary. What do they use to learn ideas such as particles, distillation or compounds in the first place? This requires a careful consideration of how learning takes place when you don't know much about the ideas you are learning about.

Brains as information processors

Our brain is responsible for processing and remembering information so learning takes place. Two main cognitive systems are used to learn: long-term memory and short-term memory (Sweller, 1988). A useful, but hugely

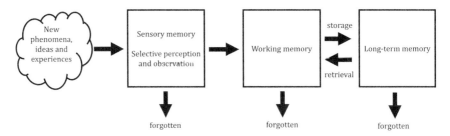

Figure 5.3 A model to help explain how humans process new information in relation to what they already know

simplistic model of this system is shown in Figure 5.3. It shows how new information is first perceived by sensory memory before being passed to working memory where it is processed with information stored in long-term memory. At all stages in this process information can be forgotten. It's worth pointing out that this model (Figure 5.3) doesn't reside in a specific part of our brain, rather, it helps us to understand *how* information is processed not *where* or *why* it's processed.

To illustrate how this system works, let's look at a classroom example where two students, Jill and Tom, are learning about dissolving – the process in which a solute (e.g. NaCl(s)) mixes with a solvent (e.g. $H_2O(l)$) to form a solution (e.g. NaCl(aq)). Or, we could equally say the process in which a substance (e.g. NaCl(s)) mixes with another substance (e.g. $H_2O(l)$) to form a solution (e.g. NaCl(aq))!

A teacher places a heaped spatula of sodium chloride (NaCl) into a large beaker. She adds hot water from the kettle and stirs. She poses the question to the class: 'what has happened to the salt'?

1. **Information is perceived by sensory memory**
   Jill: Sees a white solid being added to colourless liquid, it then disappears.
   Tom: Sees a white powder added to a hot colourless liquid from a kettle, it then disappears.
2. **Information enters working memory where it is consciously processed with information retrieved from long-term memory to make sense of ideas**
   Jill: Knows that when salt is added to water it dissolves. That's why the sea tastes salty. Jill infers that the salt has dissolved but it's still there. It's just been broken up into smaller bits she can't see.
   Tom: Knows that when you heat things they melt away and get smaller. Tom infers that the salt has melted into the hot water.

3. **New information is stored in long-term memory**

  Jill: When salt is added to water, it breaks up into smaller bits that we
       can't see anymore.

  Tom: Salt melts and disappears in hot water.

Even though Jill and Tom observed exactly the same demonstration, they
learnt very different things. Tom was looking at the kettle because it had a
mesmerising blue light and Jill was looking at Joe because he had nice hair:
both missed parts of the teacher's explanation. Neither Jill nor Tom thought
about the importance of water in the dissolving process and both focused on
subtly different aspects. For Tom, it was the 'hotness' that was the cause of
salt disappearing because he already knew that heating makes things melt.
Jill, however, understood that this process was dissolving as she already
knew that salt can dissolve in cold water, thanks to the antics of her younger
sister during her last seaside holiday.

   This example shows why learning science is not simply a corollary of what
has been taught. Learning is influenced heavily by what has been perceived
and by what is already known, not to mention other factors such as beliefs,
attitudes, classroom environments and summer holidays. However, we can
increase the chances of learning by creating tasks that minimise irrelevant
information – in this case by not using boiling water – and so focus attention
on the aspects that we want students to be thinking about. In this case, the
interaction of sodium and chloride ions with water particles. Repeating
the demonstration with oil instead of water would make this point clearer –
salt will not dissolve in oil, no matter how much you stir it.

Long-term memory: where we store information

Perhaps counterintuitively, we are going to explore the learning process by
starting at the end, that is, with long-term memory. Our long-term memory
is where we store information such as our home address, our telephone
number and the formula for salt. As the Australian education psychologist
John Sweller points out, 'we are our long-term memory' (ironically, I can't
remember where this quote came from so I am unable to provide a reference).

   We can think of long-term memory as being similar to a well-run second-
hand book shop, in that it stores all kinds of information, borrowed from
other people, and is highly organised in how this information is stored.
Related ideas, like books, are grouped together. Long-term memory differs
to a second-hand book shop, however, in two key respects. First, long-term

memory is infinite in size. As far as psychologists are aware, the ability of our minds to remember is endless (although I know it doesn't always feel this way). Second, knowledge in long-term memory is constantly being revised as we strive to make sense of ideas through discussion and reflection and it is the revision of this knowledge, values and beliefs that ultimately describes learning.

Information stored within our long-term memory is organised into mental structures called schemas. Jean Piaget used the term schema as long ago as 1923 (first published in English 1926) to explain how knowledge is organised, with related knowledge being linked together in what Ruth Ashbee (2018) helpfully refers to as a web of knowledge. I've tried to represent what a schema/web of knowledge for materials *could* look like (Figure 5.4). This is my schema and is likely to be different from your schema, but that doesn't mean either of us is wrong, although we could be. Here knowledge is organised in a hierarchical way, where concepts are linked together using prepositions (linking words) to form statements, sometimes called propositions. The concept of a compound then is understood in relation to the concept of an element: elements make up compounds and so understanding one concept is dependent on understanding many others.

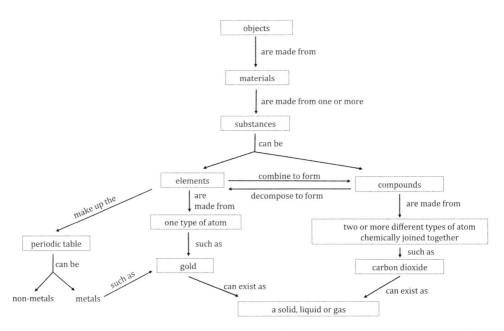

Figure 5.4 A concept map of materials and substances showing the relationships between different concepts

■ Schema formation: how we organise stored information

The formation and refinement of schemas in long-term memory is analogous to conceptual change, where students gradually revise their ideas to reflect more scientific ways of thinking. You can track these types of changes in knowledge restructuring by asking students to draw concept maps at different stages throughout the curriculum (see, e.g. Pearsall *et al.*, 1997). Piaget suggested there are two main ways in which people refine and develop their schemas: through assimilation and through accommodation (see Posner, 1982). During assimilation, the incoming information is seen to be compatible with what is already known and so the new idea simply 'slots into' the existing schema. An example of assimilation might be when a student sees a piece of gold for the first time. Gold is shiny and a solid and so can quickly be understood, or assimilated, as a metal.

Assimilation isn't always a good thing though. In our example above, Tom assimilated the observation of salt disappearing in hot water into his schema for melting. This type of incorrect assimilation is the source of many misconceptions in science. Returning to our second-hand bookshop for a moment, this is analogous to putting a cookbook into the section on gardening but being blissfully unaware of the mistake.

In contrast to assimilation, accommodation happens when the new information being acquired is seen as incompatible with currently held ideas, requiring the modification of an existing schema. Mercury is shiny like gold but unlike gold is a liquid at room temperature (mpt. −38.8 °C). Mercury is not easily assimilated into a pre-existing schema for metals, because, according to pre-existing ideas, metals are solids. There are three choices then for students to make, two of which require some mental effort.

- Option 1: They could think about something else.
- Option 2: They could compartmentalise their thinking, retaining their original schema for metals as solids whilst also creating a new schema for mercury.
- Option 3: They could modify their existing schema to recognise that metals can be liquids as well.

At times it can feel that Option 2, that is rote learning, is the best option, as it is a faster and more trouble-free strategy to adopt in the short term (Novak, 2002). However, unless there is time for students to integrate their existing knowledge with the new knowledge, students won't be able to use these ideas to inquire about the world because they are unable to see the connections between them.

Working memory

Working memory is the part of our cognitive system where new informa-tion interacts with what is already known from long-term memory. Unlike long-term memory, it is limited in capacity, being able to cope with only about seven chunks (maybe even less) of novel information at any one time, retaining ideas for only 20–30 seconds. On its own, working memory would limit humans to some pretty trivial thoughts. The good news though is that working memory is able to treat one schema (an organised grouping of knowledge that we learnt about above) as a single unit of information. This means that the limitations of working memory essentially disappear when processing information held in long-term memory because multiple schemas can be brought together when we are solving complex problems (Sweller *et al.*, 2019).

Cognitive load theory

The idea that working memory is limited when learning new information led to the concept of cognitive load theory (Sweller, 1988). You can think of cognitive load as the effort being used by working memory when con-sciously processing information. When the capacity of working memory is exceeded by the task, then learning becomes inefficient, just as a jug will overflow when it's full. You may well have experienced this effect for yourself when arriving at a foreign airport for the first time. Even the most basic of tasks, such as ordering a coffee, probably felt quite hard, but why? You have ordered plenty of coffees before. The difference is now all the information is new and you can't rely on using your ideas from long-term memory to help you out: the word for coffee, the cur-rency, whether you pay before or after are all questions that quickly con-sume you, or more precisely, overload your working memory. This is why travelling is so mentally exhausting and renders even the most basic of tasks tricky. However, if you were to visit that same airport again that year, then schemas developed from your first trip can be brought for-ward from long-term memory into your working memory, allowing you to now buy a coffee, whilst texting a friend and keeping an eye out for your luggage.

   There are three types of cognitive load that combine together to create the overall strain on our working memory (Sweller *et al.*, 2019). Let's explore each one below using examples from the science classroom.

Intrinsic load: influenced by the curriculum and by what you already know

Intrinsic load is determined by the complexity of the information you are learning about and what you know already. In our example above, it was purchasing a coffee in a foreign airport. In a science lesson this might involve learning the location of elements on the periodic table. This information doesn't have a particularly high intrinsic load associated with it because it's possible to think about each part of information in isolation. I simply need to use the periodic table to find the symbol for an element and then separately look at the group and period number. However, determining the state of a substance given its boiling point carries a much higher intrinsic load as there are a number of parts that interact when you solve this type of problem. For example, if we are asked to predict the state of bromine at room temperature, given its melting point (−7.2 °C) and boiling point (59 °C), we need to keep both these temperatures in mind whilst considering their significance to the physical state of bromine at 22 °C. This is cognitively more demanding than knowing that sodium metal is located in group 1 of the periodic table.

Extraneous load: influenced by the design of the task

When completing an activity, not only are you having to cope with the difficulty of the thing you are learning about, you also need to contend with how the new information is presented to you. This mental effort does not directly contribute to learning and is called extraneous load and is wasted effort. Staying with our example about the location of elements in the periodic table, we could minimise extraneous load by giving students a simple table to complete, possibly including a worked example to help students understand what they need to do. Table 5.1 reduces the overall complexity by allowing students to think about each aspect separately.

Or, if we were a little sadistic, we could increase extraneous load by using the same table, but this time ask students to first unscramble different key

Table 5.1 Tables can reduce the extraneous load of a task

Element	Symbol	Group	Period
Sodium	Na	1	3
Lithium			
Carbon			

Table 5.2 Increasing extraneous load does not help learning

Mneleet	Bslyom	Pruog	Oiredp
Odiums			
Hilumti			
Ocrban			

words to 'crack the code' (Table 5.2). This activity has a much higher cognitive load associated with it that isn't relevant for developing the schema we are interested in – that is, to develop an understanding of where specific elements are located on the periodic table.

Similarly, we can reduce extraneous load when trying to work out the state of a substance given its melting and boiling points using a number line (Figure 5.5). By assigning the boiling point and melting point of bromine to the scale, it reduces how many parts working memory has to think about at any one time. It's now much easier to see that bromine is a liquid at room temperature.

Germane load: influenced by the design of the task

You might think then that tasks should be designed to keep them as simple as possible so to minimise cognitive load. But there is a type of cognitive load that we want to maximise. This is germane load, and it is associated with the effort involved in developing schemas in long-term memory.

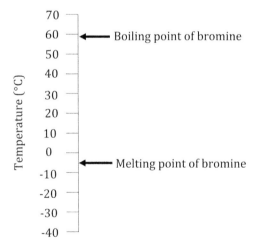

Figure 5.5 Number lines can reduce extraneous load when working out the state of a substance at different temperatures

The implications of germane load are that tasks don't want to be made easy, instead, they should encourage students to think hard about integrating the right ideas into schemas.

Take, for example, drawing particle pictures for a single substance at different states. This question could be as simple as asking students to draw a particle diagram to show the arrangement of particles in solid, liquid and gaseous water.

Students can do this activity fairly easily, in a way that involves little integration of ideas, simply by drawing little balls into boxes. However, we can tweak this activity by asking students to draw the substance in beakers rather than boxes, and then show what the substance looks like as it is poured out onto the desk (Figure 5.6).

Drawing particle pictures in this way will increase cognitive load, but in a good way, because the additional challenge gets student thinking hard about the ideas we want them to be thinking about, that is:

- that solids have a fixed shape and so don't fill their container.
- that liquids and gases take the shape of their container.
- that liquids and gases flow because particles can slide over each other, whereas solids don't.
- that gas particles will spread out as they are poured but the liquid particles stick together.

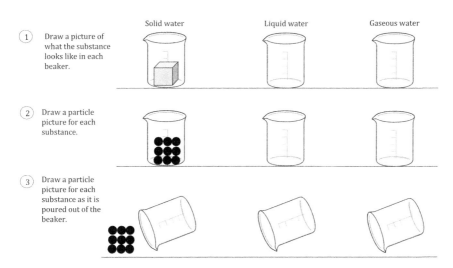

Figure 5.6 Asking students to think hard about particle arrangement and physical properties helps develop well-organised schemas

Designing tasks with cognitive load in mind

The ultimate goal of cognitive load theory has been to inform the design of instruction and classroom resources. On the surface, the particle model task in Figure 5.6 appears quite simple but it hides a number of important principles that you can use to create a variety of activities for a range of topics to maximise learning:

- it starts simply by asking students to draw the substance as they see it and so the task begins by using what students already know – the macroscopic,
- the later stages directly build on ideas from the previous stage creating a sense of progression throughout the task,
- the final stage is designed to focus student thinking about the conceptual aspects that really matter; that is, relating the macroscopic properties of the substance (volume, shape and ability to flow) to the particle model,
- within each stage there is some repetition so that students can consolidate learning through practice before moving on to the next step and
- the task encourages students to connect their thinking between the macroscopic and submicroscopic levels that we learnt about in Chapter 4.

The key then is to ensure that tasks are designed such that there is sufficient room for germane load. If the intrinsic load and the extraneous load are too high, there will be no spare capacity for students to develop the schemas we are interested in developing. However, to maximise germane load the activity needs to be carefully designed so that students are thinking about the right things. Throughout this book, we will look at ways to do this.

Summary

Any chapter that attempts to explain how human learning works needs to be accompanied by a serious disclaimer: *how the mind works is an area of research in its infancy and so ideas need to be interpreted cautiously*! There are some things that we can be reasonably confident about, though. First, learning is a process that involves the personal construction of ideas that is mediated through discussion with others. You can't learn chromatography for your students and two experts will differ in how they come to understand the same concept, yet both conceptions could still be valid. Second, knowledge stored within our long-term memory is understood in relation to other knowledge. This means that tasks should be designed to encourage students to connect one concept to another. Third, students are going to need help to develop these mental structures, so careful instruction that doesn't overload working memory is absolutely necessary and some of this instruction will need to involve conversation as students come to reject their naive ideas and accept the scientific ones. Finally, cognitive load theory does not mean that students don't have to think hard – far from it. Instead, learning materials should be designed to get students thinking hard about the *right* things – that is, the construction of fruitful schemas that generate an understanding of the ideas of science. But information processing is only part of the story for understanding learning. Humans are, after all, more than information processing machines. We have urges, insecurities, ambitions and beliefs, all of which influence the amount of effort we are willing to invest when learning science. Whenever we are thinking hard about science, we must stop thinking about something else. This requires a decision to be made, either think about science, or think about what's for dinner, the latest box set or how long left till lunch. To avoid these distractions, we need some motivation and so this is where we head next.

Bibliography

Ashbee, R. 2018. *Sentences and the web of knowledge*. Available at: https://rosalindwalker.wordpress.com/2018/10/17/sentences-and-the-web-of-knowledge/. [Accessed: 30 December 2019.]

Johnson, P. M. 2011. *Stuff and substance: Ten key practicals in chemistry*. London: Gatsby. Available at: https://www.stem.org.uk/elibrary/resource/29586. [Accessed: 15 December 2019.]

Novak, J. D. 2002. Meaningful learning: The essential factor for conceptual change in limited or inappropriate propositional hierarchies leading to empowerment of learners. *Science Education*, 86(4), pp. 548–571.

Pearsall, N. R., Skipper, J. E. J. and Mintzes, J. J. 1997. Knowledge restructuring in the life sciences: A longitudinal study of conceptual change in biology. *Science Education*, 81(2), pp. 193–215.

Piaget, J. 1926. *The language and thought of the child*. London: Kegan Paul, Trench & Trubner.

Posner, G. J., Strike, K. A., Hewson, P. W. and Gertzog, W. A. 1982. Accommodation of a scientific conception: Toward a theory of conceptual change. *Science Education*, 66(2), pp. 211–227.

Sweller, J. 1988. Cognitive load during problem solving: Effects on learning. *Cognitive Science*, 12(2), pp. 257–285.

Sweller, J., van Merriënboer, J. J. and Paas, F. 2019. Cognitive architecture and instructional design: 20 years later. *Educational Psychology Review*, 31, pp. 261–292.

Treagust, D. F. and Duit, R. 2009. Multiple perspectives of conceptual change in science and the challenges ahead. *Journal of Science and Mathematics Education in Southeast Asia*, 32(2), pp. 89–104.

Why motivation matters

Powerful idea focus

- *When substances react, atoms are rearranged and new substances form but mass is always conserved.*

British people like their candles, reportedly spending £1.9 billion on them annually (Marks, 2019). For the most part though, these candles are not being used to light up the dark, they are purchased for more emotional reasons.

Learning science is similar, in that the scientific ideas themselves aren't always that useful but they are transformative in how they allow us to see the world. And so learning scientific ideas, like watching burning candles, involves emotion too (King *et al.*, 2015; Sinatra *et al.*, 2014). Emotions are brief mental and physiological feeling states that direct our attention and guide our behaviour. In chemistry there is the joy of making a new substance or the surprise of a levitating Bunsen flame (De Carvalho, 2012). Positive emotions like these have a value in their own right in supporting well-being – an important goal considering the poor mental health of many people today – but they also play an important part in the learning process. Let's return to the burning candle for a moment to find out what this emotion feels like as we explore the powerful idea of this chapter – *when substances react, atoms are rearranged and new substances form but mass is always conserved.*

▌The burning candle

The burning candle has long been a fascination of both scientists and non-scientists and formed the topic of Michael Faraday's (1791–1867) seminal lectures, which later would become known as the 'Christmas Lectures' (Hammack and DeCoste, 2016). A candle is made from wax, which as we saw in Chapter 5 is a mixture of hydrocarbon molecules, mainly $C_{25}H_{52}$ made from carbon and hydrogen atoms. These molecules, as we will see in Chapter 8, are stuck to each other, but only loosely.

Before you read on, find a candle, take a match and light the wick and just watch it burn for a while and note down what you think happens to the wax.

Feelings of wonder

The solid wax begins to melt as energy is transferred to the wax from the flame by radiation and through the wick by conduction. As the wax melts, liquid wax gets drawn up through the wick by capillary action, just as water is soaked up by a kitchen towel.

As the wax reaches the top of the wick it turns from a liquid to a gas. This gas flows out of the wick just as gas flows out of the burners on your gas cooker at home. These wax molecules react with oxygen molecules from the air to form carbon dioxide and water (and many other products which we will ignore for now). Substantial amounts of energy are released during this reaction and soot particles glow, giving off light that gives the flame its characteristic yellow colour.

$$C_{25}H_{52} + 38O_2 \rightarrow 25CO_2 + 26H_2O$$

You can prove one of the products of combustion is water by holding a cold beaker above the flame and condensing the gas. Testing the liquid formed with blue cobalt chloride paper will turn the paper a beautiful pink, showing that water is indeed present. You can take the same beaker and breathe on it and you will also collect water. A burning candle and a human being both use combustion to do work, but for a human the fuel is sugar, whereas for a candle the fuel is wax. What's truly wonderful though is that the carbon atoms within the sugar we combust (or more accurately respire) may once have been carbon atoms in a candle, a grandma or any other object made from carbon-containing substances. Chemical reactions then recycle matter, meaning atoms can move from one substance to another but they are not created or destroyed.

Whilst chemical equations like the one above can help students to see that overall mass is conserved in chemical reactions, make sure students see what you see. For example, when the reaction above has finished, the oxygen and hydrocarbon molecules are no longer there but the equation suggests otherwise. Writing out the equation in steps and building up to the overall equation can help here.

Feelings of surprise

Now take the burning candle and blow it out. If you're quick, and hold a lit match just above it, you will be able to re-light the candle without touching it. This is surprising.

Feelings of confusion

Hold onto this surprise and it's an opportunity for confusion (D'Mello *et al.*, 2014), a time where pre-existing ideas are seen to be incompatible with new incoming information. How is it possible to light a candle without touching the wick? To resolve this confusion requires existing ideas about burning candles to be refined; what burns in a candle is not the wick, it's the vaporised wax. And what looks like smoke when you blow the candle out is actually unburnt fuel, condensing in the air as it cools which is why you are able to light it.

Feelings of fear

The products of combustion are very dangerous. Carbon dioxide is not only a greenhouse gas that absorbs infrared radiation like a black t-shirt, it is also a major cause of ocean acidification. When carbon dioxide dissolves and then reacts with ocean water it forms carbonic acid (H_2CO_3), the same acid (amongst others, e.g. phosphoric acid) that destroys our teeth if we drink too many fizzy drinks. This acid reacts with carbonate ions in the sea water that are needed by many marine organisms such as lobsters to build and maintain their shells, leading to slow growth rates and even death. And this problem is getting worse as we continue to pump out more CO_2 into our atmosphere (Figure 6.1).

Since the industrial revolution, the pH of surface ocean waters has fallen by 0.1. This may not sound much, but like the Richter scale, pH is logarithmic, meaning a change of one represents a tenfold difference. The National Oceanic and Atmospheric Administration (2013) estimates that by the end of this century oceans could be nearly 150% more acidic than today putting jobs and species at risk.

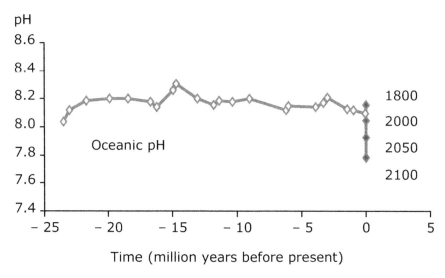

Figure 6.1 Ocean acidity over the past 25 million years and projected to 2100

Source: EEA. Credit: European Environment Agency. Attribution 2.5 Denmark (CC BY 2.5 DK).

Effects of emotion on learning

Whilst positive emotions can promote interest and confidence when learning science, positive moods are not without their problems. When we feel positive and comfortable, we can take shortcuts, perhaps driving with only one hand on the steering wheel, marking a piece of work without the mark scheme or just watching a candle burn and drifting off. Students can do the same, engaging only superficially with the ideas they are learning about. However, when we experience confusion, at least in the short term, we tend to be more careful and systematic in our thinking as we try to make sense of the discrepant or troublesome event. In our burning candle above, effortful thinking was needed to resolve the confusion that the smoke could be ignited – that is, to recognise that smoke was actually a gas and this is what burns and not the wick. This type of instructional strategy is referred to as cognitive conflict (Limón, 2001) and can be a useful way to begin to address misconceptions in the classroom as long as students are sufficiently interested in the topic and have sufficient prior knowledge to experience the conflict in the first place.

It feels then that there is a careful path to tread in the science classroom. If students are too comfortable, they will start taking shortcuts and not think deeply about what they are learning. However, if it feels too negative, anxiety and fear will run riot, causing students to withdraw to avoid attention (Plass and Kalyuga, 2019). What's needed then is to ensure students are sufficiently motivated to sustain and resolve periods of planned confusion so they experience the positive emotions of success and wonder that learning scientific ideas can bring.

Motivation

There is a close relationship between motivation and emotions, but they are not the same. Positive emotions can act to reinforce certain behaviours, but it is motivation that initiates and sustains students during the learning process (Cook and Artino, 2016).

Ryan and Deci (2000) consider two main types of motivation: intrinsic motivation (autonomous) and extrinsic motivation (controlled). Extrinsic motivation refers to doing something because it has a separate outcome, such as being awarded a sticker for neat work or a punishment for late home-work. Extrinsic motivation can get quite a bad press, but it explains why students learn many things at school, such as their times tables or physics formulae. For extrinsic motivation to work well, the purpose of the reward needs to be clearly understood so that students participate with a sense of volition. So, whilst some students may not want to wear goggles in the lab, if they understand this is for their own safety, over time they may come to internalise this idea, even though at first it was the threat of detention that made them goggle up.

Examples of extrinsic motivation in science might include:

- Praise from the teacher for packing away equipment well after a practical.
- Merit stickers from the teacher for excellent effort.
- Parental praise for doing well in a test.
- Being allowed to leave the class first when the bell goes for good behaviour.

Whilst extrinsic motivation can be effective in the short term, offering rewards to encourage people to stop smoking, lose weight or wear seat belts can cause problems later on, with 'rewarded' participants showing worse compliance than the control groups (Kohn, 1999)! Effects like these have been found within the classroom too.

Intrinsic motivation, on the other hand, is experienced when something is intrinsically interesting or enjoyable. It doesn't require an external source of validation because the task itself is exciting and engaging. This is why people risk life and limb climbing rocks and abseiling down very tall buildings, even when no one is looking. Balancing equations or making copper sulfate crystals can all be intrinsically enjoyable for some students, either because they foster feelings of competence or because they fulfil some other basic human need such as feeling connected to others and being autonomous (Ryan and Deci, 2000). What determines how intrinsically interesting activities and concepts are, to some extent, depends on what we already know, so there is a close relationship here between prior knowledge and motivation. It's hard to feel competent and autonomous when making crystals if you don't know how to light the Bunsen burner.

A motivational model for the classroom

David Palmer (2005) from the University of Newcastle in Australia has produced a helpful model to think about motivation in science classrooms that identifies three elements to focus on. I have described and illustrated each element using some specific examples below.

Selection of concepts and tasks that represent appropriate challenge

Success breeds success and so students need to meet achievable challenges to experience a sense of achievement. Success doesn't mean learning something new in every lesson, rather it means having the opportunity to face tasks of moderate difficulty and then feel successful when solving them.

> **Using appropriate challenge:** demonstrate the reaction of iron with sulfur then ask students to complete the activity below (Figure 6.2). The activity builds in difficulty, considering the macroscopic properties of the substance first and then the submicroscopic particle pictures. There is then the opportunity to use the symbols to create an equation. Between questions there is an increase in the level of challenge, but within a question there is repetition to allow for practice and the feeling of success. There is also the opportunity in the final question for students to demonstrate they understand the difference between a compound and a mixture. This brings together ideas from Chapter 5 and so draws different powerful ideas together.

Use of 'dual purpose' teaching approaches

Dual purpose teaching approaches best illustrate or assess the concept *and* motivate students. Ideas to try in your classroom include fantasy,

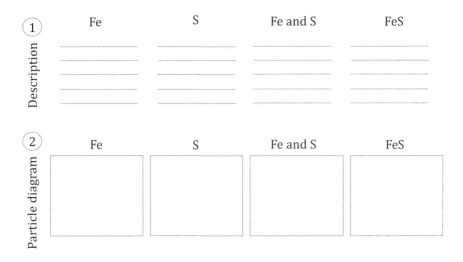

③ Write a word and chemical equation for the reaction of iron with sulfur

④ Why is iron sulfide classified as a compound and not a mixture?

Figure 6.2 Helping students understand the nature of a chemical reaction – when a new substance is made

discrepant ideas and applications to everyday life (Lepper and Hodell, 1989). What matters is that the purpose of the activity matches the goal. So, the same practical activity, such as the dehydration of sucrose using concentrated sulfuric acid, could be used to grab students' attention before instruction *or* could be used to promote the reorganisation and refinement of ideas after teaching has taken place.

>   **Fantasy:** fantasy gives students the opportunity to draw together lots of ideas in a way that generates a sense of meaning. Fantasy allows you to assess and extend understanding because you can ask questions that would never be tolerated in the 'real' world. You can pose these questions by asking '*what would happen if …?*' Make sure there is time at the end to clarify any misconceptions that might emerge from this type of task. If students are learning about decomposition reactions you could ask them how elephants brush their teeth. Then with a measuring cylinder and a dilute solution of hydrogen peroxide, food colouring, detergent and a catalyst you produce a giant tower of oxygen bubbles – which resembles a popular brand of toothpaste,

but elephant size! You can then use this demonstration as a stimulus to explore ideas such as rate of reaction, catalysis and conservation of mass.

Curiosity: curiosity can be initiated by discrepant ideas or surprises that are different to existing beliefs and ideas. Curiosity can be used to arouse confusion to refine understanding or to act as a 'hook' so students want to know more. You could ask students to predict what happens to the mass of iron wool when it is heated in air. They then observe the reaction and see that the mass increases (most will think it decreases). Students then explain their observation and during this process, refine their initial ideas, perhaps with the use of a model showing oxygen atoms combining with iron atoms.

Applications to everyday life: seeing how scientific ideas connect to everyday lives can create a reason for students wanting to know more. You could show students the effects of fizzy drinks on teeth (calcium carbonate), where elements that make up smartphones come from and why gold and not sodium is used to make jewellery.

A classroom climate that encourages positive motivational beliefs

Students should have the opportunity to experience a sense of autonomy when learning science. This could be as simple as allowing them to choose which variable to investigate or what apparatus to use when investigating factors that affect the rate of reaction. Where possible, praise effort and improvement and create a sense within the classroom of learning together.

Projects: give students the option to collaborate with other students on a project. This could involve exploring the effects and consequences of ocean acidification on organisms by carrying out an investigation that explores the effect of acid on the shells of different species of mollusc, such as mussels and clams from the local market.

Practical task with an element of choice: students work in pairs to find the perfect dilution of solutions to time, a delay of exactly 40 seconds. This could be used with either the iodine clock reaction that turns blue-black or the reaction between sodium thiosulfate and hydrochloric acid that produces yellow sulfur.

Summary

We saw in Chapter 5 that learning comes at a cost and requires some mental effort. Whenever we are thinking hard about science, we must stop thinking about something else. To engage with and persist with an activity, we need motivation. As we've seen in this chapter, motivation is both a prerequisite and a co-requisite for learning and that feelings of competence, autonomy and relatedness are important sources of intrinsic motivation for students learning science at school. There is therefore an important interplay between prior knowledge stored in long-term memory and motivation. For students to experience a sense of competence and autonomy, they need sufficient prior knowledge to be successful in the challenges and practical work that they are set. But if students aren't given the opportunity to be challenged or work with a sense of their own volition, then they won't have the opportunity to experience feelings of autonomy and competence and so motivation to learn new ideas will slowly wane. Neither will they have the drive to sustain times of confusion that play critical roles in the integration of knowledge when learning tricky concepts. Lessons then need to be planned so that they consider both the cognitive and affective aspects of learning and so the best activities are those that work hard to achieve both.

Bibliography

Cook, D. A. and Artino Jr, A. R. 2016. Motivation to learn: An overview of contemporary theories. *Medical Education*, 50(10), pp. 997–1014.

De Carvalho, R. 2012. The levitating Bunsen flame experiment. *Physics Education*, 47(5), pp. 515–518.

D'Mello, S., Lehman, B., Pekrun, R. and Graesser, A. 2014. Confusion can be beneficial for learning. *Learning and Instruction*, 29, pp. 153–170.

Hammack, W. S. and DeCoste, D. 2016. *Michael Faraday's the chemical history of a candle: With guides to lectures, teaching guides & student activities*. Urbana, Illinois: Articulate Noise Books.

King, D., Ritchie, S., Sandhu, M. and Henderson, S. 2015. Emotionally intense science activities. *International Journal of Science Education*, 37(12), pp. 1886–1914.

Kohn, A. 1999. *Punished by rewards: The trouble with gold stars, incentive plans, A's, praise, and other bribes*. Boston, MA: Houghton Mifflin Harcourt.

Lepper, M. R. and Hodell, M. 1989. Intrinsic motivation in the classroom in *Research on Motivation in Education*, Ames, C. and Ames, R. (eds.), pp. 73–105. New York: Academic Press.

Limón, M. 2001. On the cognitive conflict as an instructional strategy for conceptual change: A critical appraisal. *Learning and Instruction*, 11(4–5), pp. 357–380.

Marks, S. 2019. UK candle economy worth 1.9 bn annually. Available at: https://www.giftsandhome.net/uk-candle-economy-worth-1-9bn-annually/. [Accessed: 18 December 2019.]

National Oceanic and Atmospheric Administration. 2013. *Ocean acidification*. Available at: https://www.noaa.gov/education/resource-collections /ocean-coasts-education-resources/ocean-acidification. [Accessed 18 December 2019.]

Palmer, D. 2005. A motivational view of constructivist informed teaching. *International Journal of Science Education*, 27(15), pp. 1853–1881.

Plass, J. L. and Kalyuga, S. 2019. Four ways of considering emotion in cognitive load theory. *Educational Psychology Review*, 31, pp. 339–359.

Ryan, R. M. and Deci, E. L. 2000. Intrinsic and extrinsic motivations: Classic definitions and new directions. *Contemporary Educational Psychology*, 25(1), pp. 54–67.

Sinatra, G. M., Broughton, S. H. and Lombardi, D. O. U. G. 2014. Emotions in science education, in *International Handbook of Emotions in Education*, Pekrun, R. and Linnenbrink-Garica, L. (eds.), pp. 415–436. New York: Routledge.

Section 3

Planning lessons with thinking in mind

CHAPTER 7

Preparing to plan
Thinking about progression over time

▓ Powerful idea focus

- *The movement of charge forms electric current and causes magnetic fields.*

▓ Planning focus

- In this chapter, we begin the planning process by exploring the concept of progression. We then look at how planning starts by familiarising ourselves with the concepts we are going to teach, thinking how they connect to previous and future learning.

▓ Linking to theory

- This chapter draws on ideas from Chapter 2, which described how powerful ideas are story-like and how they should be introduced in a carefully sequenced way. This links to what we learnt about how students organise and store information according to schema theory in Chapter 5, where new knowledge is understood in relation to what is already known.

The Massachusetts Institute of Technology and Harvard are some of the most prestigious universities in the world. Yet, back in the 1990s, graduates from these elite institutions struggled to light a bulb when given a battery and a single wire (Harvard-Smithsonian Center for Astrophysics, 1997). It seems, even among these Ivy League graduates, their understanding of electricity had diverged from what the school curriculum had intended them to learn about circuits needing to be complete for current to flow (Model B, Figure 7.1).

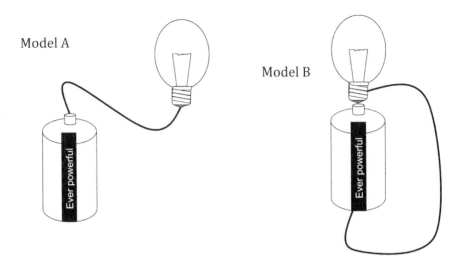

Figure 7.1 Can you light the bulb when given a battery and single wire?

Note: Many pupils (and graduates) see model A as correct, where the battery is a source and the bulb a consumer. However, this circuit is not complete. Model B shows the correct idea.

This mismatch, between the intended curriculum – what we want students to learn about circuits – and what they actually learn is somewhat inevitable when you think about how scientific ideas are learnt. That is, they are personal understandings of an abstract world that can't be seen. However, we can try to make this process of constructing a scientific understanding as smooth as possible by planning for progression in understanding.

Progression

Progression is the personal journey that all learners embark upon as they develop an understanding of a concept or the ability to do something (Braund, 2008). Progression in science describes how we move from a naive way of thinking to a more scientifically acceptable idea. This involves progression in understanding scientific ideas or the nature of science. So, students may start off thinking that electricity is used up as it travels from the plug along one wire to the television set but will later learn that one wire is actually two (or three) and that circuits must be complete for current to flow. As their understanding develops further, they will find out that current – that is, the flow of negative charge – moves around a circuit when a potential difference (voltage) is applied.

We will come back to potential difference in a minute, but for now I hope you can see how understanding electricity mirrors what we learnt about powerful ideas being a story in Chapter 2. The order in which ideas are introduced matters if students are going to make sense of the science story and so narrative plays an important role in communicating scientific ideas to students (Avraamidou and Osborne, 2009). Planning for progression then involves sequencing knowledge in the curriculum to give students the best possible chance of learning these new ideas in a *meaningful way* – that is, relating them to what they already know so that they come to appreciate the significance of science, not just to society, but also to their own lives.

Meaningful learning

David Ausubel (1963) made the distinction between rote and meaningful learning back in the 1960s. Rote learning is where new knowledge is arbitrarily learnt and remains disconnected from what is already known. Students may learn and be able to chant the definition that 'current is the rate of flow of charge' around a circuit but will look at you blankly if you then ask them what charge is or how it flows. Such disorganised knowledge is difficult to use, although it can still pass examinations.

Meaningful learning, in contrast, involves the careful integration of new knowledge into pre-existing ideas, resulting in the well-organised knowledge structures (schemas) that we learnt about in Chapter 5. In the example above, not only can I define what current is and know that in wires it is negative charges that move, I know the negative charge in a wire is carried by electrons and that these electrons come from the metal atoms that make up electrical wires. Together these ideas form a chain of understanding (Figure 7.2) that students use to reason with.

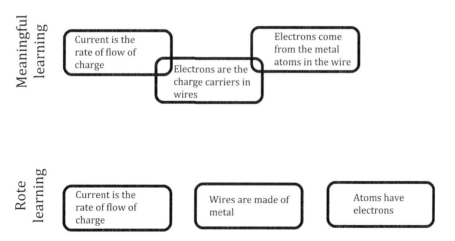

Figure 7.2 In meaningful learning, knowledge can be connected together like links in a chain

Planning for progression then requires us to think carefully about the order in which ideas are introduced to give students the best chance of connecting these ideas in a logical way. As we explore the powerful idea of this chapter below – that is, *the movement of charge forms electric current and causes magnetic fields* – we will see that this often starts with the easier familiar ideas and moves to the abstract harder ones, but this is not always the case.

Sequencing knowledge – forming the science story

We saw in the previous paragraphs how charge flows around an electrical circuit in the wires – but why? Why do charges move at all? To answer this question we need to remind ourselves of what we learnt in Chapter 5 about matter being made from atoms and that atoms are built from two basic charges, positive and negative. When charged objects are brought together, they exert a force on each other, even when they are not touching, and this causes charges to move. We can represent this below by looking at how positively and negatively charged spheres behave when hanging from a rope (Figure 7.3). Spheres with the same charge have a tendency to move apart from each other, whereas spheres with opposite charges attract. These spheres could represent electrons, protons, ions or even balloons.

We have already seen how starting with what students know can be helpful when planning for progression. For that reason, let's use some everyday objects – a balloon made from rubber and a jumper made from wool – to explore further how charges behave before we return to our original question of why charges move in electrical wires.

Our starting point is to recognise that the balloon and the jumper have no overall charge. This is because both objects are made from atoms, and atoms have equal numbers of positive protons and negative electrons. As we rub these materials together, however, friction shaves off electrons on the outside of atoms. Because different materials have a different tendency to attract

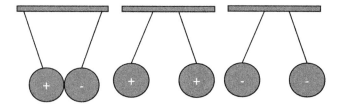

Figure 7.3 Opposite charges attract and the same charges repel

electrons, there is an overall direction in which electrons get transferred. In the case of the balloon and the jumper, the negative electrons move from the jumper to the balloon. This causes the balloon to become negatively charged and the jumper to become positively charged. It's worth pointing out that overall, charge has been conserved here, it's just moved around a bit. These oppositely charged objects will now attract each other, like the spheres in Figure 7.3, and so the balloon sticks to the jumper (or does the jumper stick to the balloon?). This effect is not just useful for party tricks, it also explains interesting things such as how positively charged bumble bees attract and carry negatively charged pollen or why clothes from the tumble dryer stick together.

If we create a big enough difference in charge between two objects, then something amazing happens – we can cause charged particles to move. Dragging your feet along a wool carpet causes electrons from the carpet to be deposited onto the soles of your shoes. These electrons can't then get off because rubber is an insulator. Over time these electrons repel each other, eventually creating a build-up of negative charge over your entire body.

Now, if your hand approaches a metal door handle you're in for a nasty surprise (Figure 7.4). The electrons in your hand repel the electrons in the door handle causing an area of the handle to become positive (metals are better conductors than flesh, therefore it is their electrons that move away and not yours). A voltage, or potential *difference*, has now been created between your negative hand and the positive handle, causing electrons to suddenly flow

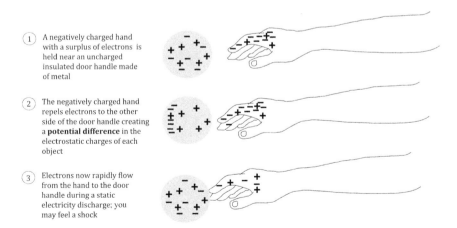

1. A negatively charged hand with a surplus of electrons is held near an uncharged insulated door handle made of metal

2. The negatively charged hand repels electrons to the other side of the door handle creating a **potential difference** in the electrostatic charges of each object

3. Electrons now rapidly flow from the hand to the door handle during a static electricity discharge; you may feel a shock

Figure 7.4 Why you may experience a small shock when opening a door

Credit: https://pixabay.com/vectors/arm-hand-wrist-human-anatomy-body-153258/.

from your hand to the handle – OUCH! You have just experienced a static discharge which was responsible for the shock you experienced. The scene has now been set to introduce electrical circuits and so I hope the story is progressing nicely.

Current electricity

Current electricity that flows around an electrical circuit, unlike static electricity, is always on the move because the battery provides a constant source of potential difference. When a battery is connected, a difference in electric potential is introduced (between the positive and negative terminals), causing negative electrons to flow around the circuit in one continuous chain, away from the negative terminal from which they are repelled. Batteries then are just the site of chemical reactions that involve the loss of electrons from one metal (the anode) and the gain of electrons by another metal (the cathode) and the wires connect the two.

In an electrical circuit, energy from the battery is gradually transferred to the surroundings by the action of various components in the circuit. In the case of the light bulb that we met at the start of this chapter, this would be through heating and lighting. This is why you will need to replace the battery in your TV remote control – not because energy has been used up in the circuit (the law of conservation of energy tells us this can't happen), but because energy has been transferred out of it. We will take a closer look at how energy is stored and transferred in Chapter 12.

Making connections

As we saw above, meaningful learning requires students to make valid connections between ideas. When we plan for progression, we need to make these connections, or relationships, explicit. So, we will want to connect what we learnt about static electricity to an electrical circuit by identifying equivalent or so-called analogous parts. We need to point out that both forms of electricity rely on negative charges being able to move, either to form the current in current electricity or to form the charge imbalance in static electricity. We should show how the negative terminal of a battery is analogous to the negatively charged hand with an excess of electrons, and that the positive terminus of the battery is analogous to the positively charged area of the door handle. Between both the terminals of the battery and between the hand and the handle, there is a potential difference and it is this difference that enables charge to move.

We also need to prevent false connections from being made. There are clear differences between these two types of electricity. In electrical circuits, for example, charges continue to move because the battery provides a constant source of potential difference, whereas in static electricity, once charges move to restore the charge balance, the potential difference has gone.

Planning for progression at scale

So far we have seen how sequencing knowledge gives students the best chance of connecting new ideas to what they already know. In the example above, we started with static electricity and described the progression towards understanding current electricity. There are, though, many other valid ways of sequencing progression in understanding electricity – what's important is that you understand the rationale behind the journey you are teaching (Figure 7.5).

Preparing to teach a unit

For most teachers the process of planning begins by internalising and understanding the unit of work that has already been planned for you. Units of work, often called schemes of learning or medium-term plans, are important documents for teachers because they define the key knowledge and skills students need to learn, along with some suggested activities.

In short, this means that students can move from teacher to teacher and year to year and still be part of the same curriculum. As well as providing the route, units of work help to create a sense within the department of what it means to get better in science because it charts the progression

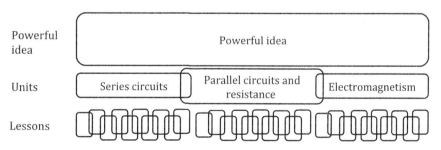

Figure 7.5 Thinking about progression using different-sized chunks of curriculum time

in understanding the scientific ideas. However, a unit of work is only the starting point, it needs to be transformed into something that students can experience. For this to be possible, it's important that we internalise and understand how progression was planned for within the unit we are about to teach.

To illustrate how this might be possible when preparing to teach a unit, let's look at an example involving a unit on electromagnetism for students aged 13+. As you develop your confidence in planning, you will adapt these steps to fit your own way of doing things but I hope they provide a useful starting point.

Step 1: identify what students are learning in the unit

Planning starts by identifying the concepts and skills that you want students to learn within the unit and to identify what's significant or powerful about the ideas that students will be learning about. These ideas can be identified by consulting the unit of work and can then be transferred onto a concept map (Novak, 1990) that we saw in Chapter 5 to help you identify relationships between the different concepts you will be teaching. This process familiarises your with the subject matter and, alongside a good textbook, can be an important step in developing subject knowledge. Remember, you are not creating the curriculum from scratch at this point. Rather, you are making sure that you understand the relationships between the ideas in the planning documents that have been created for you and that you have an understanding of why students are learning these ideas in the first place.

Below are some key questions to help you build your concept map and further help can be found here: https://msu.edu/~luckie/ctools/. Don't worry too much about getting the 'right' concept map – this doesn't exist, so just put down the ideas that you will be teaching and connect them together; it is the process, not the product, that is important here and it is absolutely fine for it to look messy.

a. What is powerful about the scientific ideas you are teaching? What new perspective do they bring to students' lives? What is awesome or disturbing about these ideas?
    ● Write this at the top of your concept map to focus your thinking.
b. What are the concepts and skills taught in this unit that underpin this powerful way of thinking?
    ● Write these down as a list first. You might find it helpful to consult a textbook at this stage to develop subject knowledge.

- Now arrange these concepts hierarchically on your map – the most general (largest) concepts at the top. This is hard but the thinking will help you to develop your understanding.
c. What are the relationships between the concepts?
  - Join related concepts with straight lines and add labels (prepositions) to the lines to describe the relationship.
  - All concepts are related, so just try to pick out the most important relationships.

Step 2: zoom out to gain perspective

To understand London, we need to recognise that it's a city located in the South of England, which is located in the United Kingdom which sits within the continent of Europe. Planning then involves a similar process of orientation, where you identify where the concepts you are about to teach sit in relation to what students learnt before and what will come next. This will probably involve looking at other units from the same powerful idea and consulting long-term plans (Table 7.1). You can then add this information onto the concept map you created in step 1 by using a different colour to distinguish prior from new learning. At this point, you are not defining these concepts in detail but rather identifying what concepts have been taught already.

Table 7.1 Look at progression across years by consulting long-term plans

Prior ideas	Age 11	Age 12	Age 13	Powerful idea
Appliances Conductor Insulator Series circuits Switches Symbols Devices Cell number and brightness Attraction and repulsion of magnets Magnetic materials Poles Uses	• Static electricity • Making circuits • Electric current and voltage in series circuits	• Resistance, $V = IR$ • Electric current and voltage in parallel circuits	• Magnetic fields and electromagnets • Generating electricity • Paying for electricity	The movement of charge forms electric current and causes magnetic fields

Table shows unit names. Adapted from Best Evidence Science Teaching resources (https://www.stem.org.uk/best-evidence-science-teaching).

Step 3: sequence the concepts and skills and allocate them to specific lessons

This may well have been done for you in the unit plan and if so, great! Check to see that you understand the logic of how the knowledge and skills are sequenced.

a. What is the science story for each lesson and the overall unit?
b. What are the particularly tricky ideas that might need more time?
c. Is there sufficient time to teach the concepts and carry out any practical work?
d. How is the learning assessed half-way and at the end?
e. Where is homework set?

If this sequencing hasn't been done you will want to now take the concepts and skills identified on your concept map and sequence these carefully, finally allocating them to individual lessons – it's best do to this type of sequencing in collaboration with another colleague as discussion can be really useful to help justify why one idea should come before another. An example is below.

> **Lesson one:** repulsion, attraction, magnetic materials, poles, Earth
> **Lesson two:** permanent and induced magnetism
> **Lesson three:** magnetic field lines – strength and direction, north-seeking pole and south-seeking pole, concept of force, effect of distance
> **Lesson four:** making an electromagnet, current and magnetic fields
> **Lesson five:** changing the strength of an electromagnet using solenoids and an iron core, uses
> **Lesson six:** making the strongest electromagnet, correlation, variables
> **Lesson seven:** explaining how the electric bell works

Step 4: develop your teacher knowledge

At this point you will probably want to go and do some further reading to develop your knowledge before planning individual lessons. There are two main forms of knowledge that teachers spend time developing. The first is content knowledge, often referred to as subject knowledge, for example knowing that current is measured using an ammeter connected in series and the unit of electric current is the ampere (or amp, A). This is perhaps best developed through the use of a good textbook, answering past examination questions, completing multiple-choice questions or speaking with a colleague.

Teachers are not unique in having content knowledge, though – all science graduates do. What is unique to teaching is something called pedagogical content knowledge (Kind and Chan, 2019), commonly referred to as PCK. PCK allows teachers to ask and answer some important questions:

- What is the best way for me to teach …?
- How do I explore what my students already know about…?
- What misconceptions are students likely to have about …?
- Which is the best model to teach students about …?
- What do students find difficult about …?

PCK is developed over a lifetime, by engaging with research, reading books, working with colleagues, training and through reflecting on classroom practice. As PCK develops, you will acquire an increasingly rich, yet specialised understanding of both the content that is being taught *and* how best to teach it, including what resources to select (Lee and Luft, 2008). It is this PCK that distinguishes the professional science teacher from the scientist and provides the conceptual tools through which you can teach and plan lessons so that *all* students have the opportunity to learn scientific ideas in a meaningful way.

The subject associations below provide a number of excellent resources to support you in developing both subject and pedagogical content knowledge.

The Royal Society of Chemistry (https://www.rsc.org/)
The Royal Society of Biology (https://www.rsb.org.uk/)
The Institute of Physics (http://www.iop.org/)
The Association for Science Education (https://www.ase.org.uk/)

Summary

Whilst the overriding temptation is to get straight into planning individual lessons, it's important to stand back and gain an overview to see where the unit or lesson you are teaching sits in relation to previous and future learning. In this chapter, we have explored the concept of progression by thinking about how students might come to understand current electricity by starting with static electricity. This built upon what students know about atoms being composed of two basic charges. We have seen how planning starts with getting a sense of the unit you are going to teach, by creating a concept map that identifies the key concepts, the relationships between them and why they are worth learning about. Then, by consulting long-term plans, it is possible to zoom out and gain a greater understanding of how the new learning connects to previous and future learning. Whilst hopefully you will be teaching from a curriculum that has most of the sequencing done for you, it's crucial that you still take time to understand how ideas are sequenced and connected together. It is from this model of progression that you will plan your explanations and know how to extend those students who are finding it easy and support those who are finding it hard. It's time now to consider how we might use the information from this chapter to plan for progression within an individual lesson.

Bibliography

Ausubel, D. P. 1963. *The psychology of meaningful verbal learning*. New York: Grune & Stratton.

Avraamidou, L. and Osborne, J. 2009. The role of narrative in communicating science. *International Journal of Science Education*, 31(12), pp. 1683–1707.

Braund, M. 2008. *Starting science … again?: Making progress in science learning*. London: Sage.

Harvard-Smithsonian Center for Astrophysics. 1997. *Minds of their our own*. Available at: https://www.learner.org/series/minds-of-our-own/1-can-we-believe-our-eyes/. [Accessed: 1 January 2020.]

Kind, V. and Chan, K. K. 2019. Resolving the amalgam: Connecting pedagogical content knowledge, content knowledge and pedagogical knowledge. *International Journal of Science Education*, 41(7), pp. 964–978.

Lee, E. and Luft, J. A. 2008. Experienced secondary science teachers' representation of pedagogical content knowledge. *International Journal of Science Education*, 30(10), pp. 1343–1363.

Novak, J. D. 1990. Concept mapping: A useful tool for science education. *Journal of Research in Science Teaching*, 27(10), pp. 937–949.

CHAPTER 8

Planning what to teach in the lesson

Powerful ideas focus

- *Substances are held together by electrostatic forces of attraction.*

Planning focus

- In this chapter we look at how to determine, in detail, what it is we want students to be able to do and know by the end of a lesson.

Linking to theory

- This chapter draws on ideas from Chapter 5 on cognitive architecture in that new learning needs to consider (i) what students already know and (ii) not overload students with too many new ideas. It also links to Chapter 3 because what we want students to learn also needs to help them to understand the nature of science.

Keeping the idea of progression in your mind from Chapter 7, I want you to close your eyes for a moment and imagine what a great science lesson might sound like if it were a piece of music.

It might start with a basic melody, a collection of notes that repeats and slowly develops throughout the piece. There would probably be clear sections, marking the beginning, middle and end. The melody may initially be played by one instrument but later will be joined by an orchestra of sounds, which get progressively more difficult to recognise but more interesting to listen to. As the music builds, its potential to move us would become increasingly real.

Composing music and planning science lessons are perhaps more similar than we think. In both cases there is the illusion that we are creating something from nothing, when in fact we are piecing together different parts in a way that (hopefully!) creates a masterpiece, or at least something that is pleasurable to experience! We also need to imagine and visualise what the plan might 'sound like', so that it can be refined and developed before it is performed.

▊ The three aspects to consider when planning a science lesson

Whether we are planning for learning in science or composing a symphony, it can be helpful to think about three distinct aspects of the planning process – the 'what', 'how' and 'form'.

> **The what:** this is what you want students to learn. In music this is the melody, a catchy phrase that you may remember after listening to a song. In teaching, this is what you want students to know and be able to do – for the most part this is knowledge of concepts or the nature of science.

An example: *know that molecules, such as water, are 'sticky' because they have electrostatic forces of attraction between them.*

> **The how:** this is about how the 'what' is transformed into something that can be truly experienced. In music this is the harmony, and works to support the melody, contributing towards the depth and colour of a piece. In a science lesson, this is the activity that takes the 'what' and turns it into something that can be experienced by the students and so involves the knowledge being transformed by the teacher in some way. This might involve using a model, a demonstration or teacher explanation.

An example: *staying with the example above, we might show students a video of astronauts playing around and popping a water balloon in space and seeing that the water sticks together. Or we might try to see how many drops of water we can fit onto a two pence piece. This would then be followed up with a teacher explanation and perhaps some questions.*

> **The form:** this is the sectioning that helps organise a lesson into clear phases that are informed by the science of learning that we met in Chapters 4–6. In music, form is used to organise the piece into

a beginning, middle or end and so gives a piece a sense of a story. In a science lesson, the form is underpinned by theories of cognitive psychology and so gives students the best chance of not only learning the information but also remembering it.

Example: *we may start our lesson by quizzing students to help them remember previous learning on the structure of atoms.*

A well-planned lesson then ensures that the 'how' and the 'form' work hard to support students to learn the 'what'. In this chapter, thinking about how we define 'what' we want students to learn will be our focus. We will focus further on the 'how' and the 'form' in Chapter 9.

What do you want students to learn?

In Chapter 7, we looked at how medium-term planning can help to prepare us to teach a unit.

Through creating concept maps we develop an overview of the concepts being taught and understand how these ideas fit in relation to previous and future learning. This level of detail is not sufficient to plan a lesson, though – it needs to be unpacked further to reveal the precise knowledge and skills that we want students to learn.

To illustrate what I mean, let's take the powerful idea of this chapter – that is, *substances are held together by electrostatic forces of attraction* – and explore how salt, a substance that was once used to pay Roman legionaries their salary or *salarium*, holds together.

Sodium chloride, better known as table salt, is quite a wonderful compound. It can be made from a toxic gas (chlorine) and a reactive metal (sodium), yet this substance can be sprinkled on your chips with no immediate harm, although long-term your blood pressure may suffer.

$$2Na(s) + Cl_2(g) \rightarrow 2NaCl(s)$$

Much of the beauty of sodium chlorine, sorry, chloride (each chlorine atom has gained an electron, so the name changes) stems from the electrostatic forces of attraction that hold the beautiful cuboid crystals together. If we could zoom into these crystals, we would see a vast repeating structure – an ionic lattice, where positive sodium ions stack on top of negative chloride ions in an alternating pattern to minimise repulsive forces and to maximise attractive forces.

Within the lattice, each sodium ion is surrounded by six chloride ions and vice versa (Figure 8.1). The many and relatively strong forces of attraction

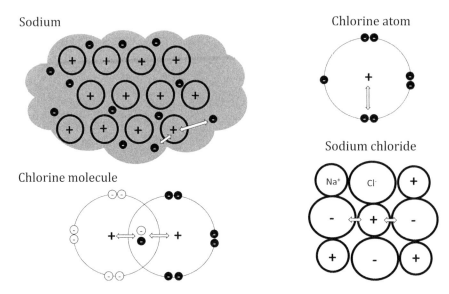

Figure 8.1 Sodium, chlorine and sodium chloride, like all substances, are held together by electrostatic forces of attraction; the arrows show some of the forces of attraction that exist; there are also forces of repulsion in all substances that are not shown

between these oppositely charged ions, commonly referred to as ionic bonds, but equally could be called ionic forces, account for salt's high melting point and overall stability. Quite a bit of energy must be transferred to pull these ions apart when it melts. But sodium chloride has a softer side. Take a hammer and strike a crystal hard and it will break up into a million or so pieces. This is because salt is brittle, and when a force is applied the lattice structure breaks apart as the repulsive forces within the lattice win over.

Amazingly, the same strong ionic bonds that are broken when sodium chloride melts can be broken by dissolving salt in water. Here, the sodium and chloride ions form electrostatic forces of attraction with the water molecules instead of each other. The energy released during this process 'pays' for the breaking of the ion-ion bonds within the lattice. This though, as we saw in Chapter 5, won't happen if you place salt in oil – oil is a mixture of covalent molecules that refuse to get intimate with ions, instead they stick to themselves, literarily, through electrostatic forces of attraction between molecules.

Objectives and outcomes

The question then for anyone teaching about ionic bonding is what do we want students to learn? Whilst we may have written down 'properties' or 'lattice structure' when creating our concept map in preparation to teach a unit on ionic bonding, these are only general ideas and are insufficient to plan a lesson – there is simply too much ambiguity in exactly what knowledge and skills we want students to learn.

Historically, lesson plans and curriculum documents have tended to clarify learning intentions by using objectives (what we are learning about) or outcomes (what I'm looking for). However, an objective such as 'be able to explain the properties of ionic compounds' is only useful if the ideas that sit beneath this objective are also defined. For example, are you teaching solubility, brittleness or conductivity, or all three? And if you are teaching about solubility of ionic substances, how are you describing the interactions between the ions and the water molecules as all of the following are possible: forces, electrostatic attractions or ion-dipole interactions?

Experienced teachers look at general objectives like these and are clear on what they need to teach as they have taught these lessons before. However, for more inexperienced teachers, this can be much more difficult and so it's worth spending some time not only defining what you want students to learn, but also then being explicit about exactly what this learning looks like in the classroom.

Objectives as key questions

Instead of phrasing learning intentions as objectives, for example, be able to explain the properties of ionic substances, they can be phrased as key questions. Key questions are helpful because they kill two (metaphorical) birds with one stone. Not only do they serve as an objective but by then writing out the expected answer they help us to define exactly what students need to know to be successful. It is the expected answer then and not the key question that informs how we plan the activities for the lesson and what prior knowledge students need. Let's take a closer look at what this may involve for a lesson where students aged 15 are learning about the properties of metals, another group of substances that are held together thanks to electrostatic forces of attraction between oppositely charged particles.

Gold, like salt, is a solid at room temperature and throughout history has been an equally prized substance to have. Unlike salt, gold will conduct electricity as a solid, which according to what we learnt in Chapter 7 means charges must be able to move – indeed they can. Unlike salt, gold can be hammered into shape, without shattering, to make prized objects like Olympic gold medals, but unlike salt won't dissolve in water, although it will dissolve in a mixture of hydrochloric and nitric acid commonly referred to as aqua regia. Incidentally, this is how two Nobel Prize gold medals were successfully hidden at Bohr's Institute of Theoretical Physics in Copenhagen during the Nazi occupation of Denmark during World War II. Thankfully, gold can be precipitated out of acid and so the prize medals were able to be recast following the war and returned to their rightful owners (Gignac, n.d.).

To understand why metals have different properties to ionic compounds let's plan a lesson on metallic bonding. We will assume that students have already learnt about the bonding and structure of ionic compounds.

The overall goal for the lesson is to explain why metals can conduct electricity whilst other substances, such as solid sodium chloride, can't.

The three key questions and expected answers I've selected for this lesson are as follows:

1. **Name three properties of metals and give a use for each**
   Good conductor of electricity – for making electrical wires
   High melting points (except mercury –38.8 °C) – for making jet engines
   Malleable — can be hammered into shape – for making jewellery
2. **Draw a labelled diagram of a metallic lattice and describe the differences between a metallic bond and an ionic bond**
   A metallic bond is the electrostatic attraction between the positive metal cations and the delocalised electrons (Figure 8.2). It is similar to an ionic bond because it is an electrostatic force of attraction; however, it is different to an ionic bond because the attraction is between positive ions and the sea of delocalised electrons and not between positive ions and negative ions.

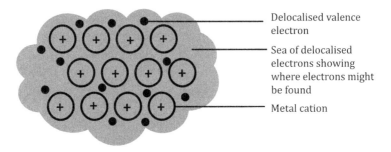

Delocalised valence electron

Sea of delocalised electrons showing where electrons might be found

Metal cation

Figure 8.2 A model of a metallic lattice

3. **Explain why electrical wires are made from copper metal and not solid sodium chloride**

    In copper metal the outer shells of adjacent atoms overlap. This means the valence electrons are not associated with any one atom and are delocalised, meaning they are free to drift throughout the lattice. When a potential difference is applied to the metal, the delocalised electrons can move throughout the lattice structure completing the circuit. Solid sodium chloride has no ions or electrons that are free to move and complete the circuit and so cannot conduct electricity. Sodium chloride is also brittle and so cannot be shaped into wires.

Why are key questions so important?

Hopefully you can see why writing the expected answers is so important. Not only does the process of writing the expected answer help to decide on the idea students are learning about, it also defines the scientific terminology we want them to use. Below I have explained some of the choices I made when defining the expected answers above.

- In key question one, the term malleable was defined to make sure I am clear on how I want students to understand this term.
- In key question two, the term cation and not positive ion was used because I want to prepare students for the types of question they may get asked in examinations. I also want to prepare them for studying electrolysis where they will meet this term, alongside cathode.
- By asking students what's the difference between ionic bonds and metallic bonds in key question two, I hope it will help students to recognise the differences (and similarities) between ionic and metallic bonding.
- In key question three, I wanted to refer to ideas about potential difference and complete circuits so that students have an opportunity to bring together knowledge from different powerful ideas.

Key questions and progression

Key questions are not just about defining the knowledge students need to learn, they are also about defining the progression of ideas within the lesson. In our example above, we started with familiar, observable features such as physical properties of metals that students are likely to know something about. The second key question introduced the idea of a metallic lattice

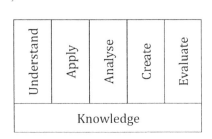

Figure 8.3 Just some of the different ways that knowledge can be used by students

Source: Inspired by Krathwohl (2002).

which was used to explain the bonding and structure of metals introduced in the first question. The final question asks students to explain conductivity using the model from key question two.

It can sometimes be useful when writing key questions to think about the different ways in which the same knowledge can be used (Krathwohl, 2002). For example, we can take the idea of a metallic lattice and ask students to do lots of different things with this knowledge (Figure 8.3).

- **Understand:** make sense of information, perhaps by explaining it to someone else.
- **Apply:** use the knowledge in a new context, for example explain why metals are good thermal conductors.
- **Analyse:** identify relationships, for example compare boiling points of different metals and spot patterns.
- **Create:** put ideas together to make something new, for example create your own model for a metallic lattice or make a prediction about the melting point of magnesium compared to sodium.
- **Evaluate:** make a judgement, for example consider the strengths and limitations of a specific model.

It's wrong to see any of these five processes in Figure 8.3 as existing within a hierarchy – that is, evaluation is not necessarily harder or easier than analysis, although it could be. Neither should these be seen as generic skills that can be easily transferred between different topics or subjects because they all rely on having subject knowledge. Rather, thinking about these different processes when framing key questions is important because they form a valuable part of the learning process itself. Not only do a variety of tasks, such as creating models or analysing data, make learning enjoyable, students will refine and extend what they know about a concept as they create, analyse and apply. As we will see in Chapter 15, they will also reveal quite a bit of what they know as they do this.

Key questions for who?

You may see learning objectives/outcomes being phrased in ways that consider who is learning them. Taking our key questions above we could say that:

1. **All** students can name three properties of metals and give examples.
2. **Most** can draw a labelled diagram of a metallic lattice and describe the differences between a metallic bond and an ionic bond.
3. **Some** can explain why electrical wires are made from copper metal and not solid sodium chloride.

On the surface, this seems like quite a sensible thing to do and recognises that some ideas are harder to learn than others. The problem with starting out in this way, though, is that it underestimates what students are capable of and overstates what we know about individual students before the lesson begins. On the whole, the human brain is pretty impressive and what limits us is not our neurology but the expectations we and others have of it.

To illustrate what I mean, let's return to the 1960s where a small study took place in a Californian elementary school. Teachers were told that a number of students in their class had been identified through a test as being late intellectual bloomers. Unbeknown to the teachers, no such test had been administered and the late bloomers had simply been allocated at random (Rosenthal and Jacobson, 1968). However, despite being chosen at random, these 'late bloomers' went on to make better than expected progress compared to the control group. This led to the so-called Pygmalion effect – that is, that changes in teacher expectations have an effect on what students go on to achieve. Whilst there are issues with the methodology used at the time, there have been a number of studies since that show that teacher expectancy matters (Weinstein, 2018). So, if you expect that only some students are going to reach key question three, then you may well be proved right, but perhaps for the wrong reasons.

The second problem with assigning objectives to all, some and most is it treats learning as all or nothing. Instead it is likely that students will learn different things. Some students might learn all of key question one, some will understand some aspects of key question 2 and all of key question 3. Instead, let's plan for all students to understand all questions but scaffold the lesson appropriately so as many students as possible will get there.

Summary

Planning great science lessons, like composing great music, is more a process of piecing together than creating something totally new. In this chapter, we have seen how planning a lesson involves thinking about three distinct parts – the 'what', 'how' and 'form'. The planning process starts with defining what we want students to learn, and this is informed by medium-term planning where we have considered what students already know and what students will learn in the future. By framing learning intentions as key questions, and then writing the expected answers, it encourages us to think carefully about the specific terminology students are learning about and how ideas build throughout the lesson. In this case, starting with familiar ideas about properties of metals and moving to more abstract ideas that explain these properties. Once the 'what' has been defined and expected answers clarified, we can consider how students should experience these scientific ideas in the best possible way. This requires us to look in a little more detail at the remaining two aspects of lesson planning, the 'how' and the 'form'.

Bibliography

Gignac, S. n.d. *The invisible prize*. Available at: https://www.aip.org/history-programs/news/invisible-prize. [Accessed: 30 December 2019.]

Krathwohl, D. R. 2002. A revision of Bloom's taxonomy: An overview. *Theory into Practice*, 41(4), pp. 212–218.

Rosenthal, R. and Jacobson, L. 1968. Pygmalion in the classroom. *The Urban Review*, 3(1), pp. 16–20.

Weinstein, R. S. 2018. Pygmalion at 50: Harnessing its power and application in schooling. *Educational Research and Evaluation*, 24(3–5), pp. 346–365.

CHAPTER 9

Planning how to teach it

Powerful ideas focus

- *Substances are held together by electrostatic forces of attraction.*

Planning focus

- In this chapter, we look at how to take what it is we want students to know – that is, the key questions and their expected answers – and then transform this knowledge into something that students can experience, reason with and learn from.

Linking to theory

- When creating activities, we need to take account of how *and* why students learn ideas in science lessons. This chapter therefore draws on ideas from Chapters 5–6 on cognitive science and motivation.

There is a wonderful ancient parable involving six blind men touching an elephant to work out what it is. Each man describes only the part of the elephant that he touches and so learns something different about what an elephant is. One man touches the side of the elephant thinking it is a wall. Another man touches the trunk – he thinks he's touched a snake.

This parable is quite a useful analogy for thinking about how students experience and learn science at school (Figure 9.1). That is, they can't see the scientific ideas directly and so must come to understand what they are

Figure 9.1 We can't see scientific concepts and so must come to understand them by making inferences from what we hear, taste, see and touch

Source: Drawing by Rory Stormonth Darling.

by interacting with various representations such as models, animations and practical demonstrations. So, whilst we may have taken time to painstakingly define what it is we want students to learn using key questions, we now need to take these abstract ideas and portray them in more tangible ways by drawing on examples from the real world. Often these examples involve making comparisons to what students already know. So, just as a blind man may make sense of a trunk by relating it to a snake, students may come to understand the abstract idea of polymerisation by relating this process to making decorative paper chains. We might also choose to demonstrate the synthesis of a polymer by making nylon in the lab and pull out the 'nylon rope' as it forms in the reaction mixture. In other words, students don't learn scientific ideas directly, they learn them through a more indirect route involving representations and manifestations which are accessible to their senses (Lederman *et al.*, 2014).

This means students need different and progressively more sophisticated ways of interacting with scientific ideas throughout their time at school, and just like the blind men and the elephant, they will need time to piece together these different interactions to make sense of the whole.

To illustrate what I mean, imagine that you wanted to teach someone what a can of Coca-Cola was. You could choose to start by simply giving them information like this:

A can of coke is a cylindrical metal container holding an aqueous solution of carbonated sugar and caramel.

You may then ask students to copy this definition into their books and then provide a few exercises to see if they had learnt it, like this:

1. A can of c_____ is a cylindrical m_____ container holding an _____ solution of carbonated sugar and caramel.
2. What shape is a can of coke?
3. What does the can hold?

But this type of activity does not teach what a can of coke is, rather it teaches the information about what a can of coke is. To understand a can of coke, you need to see one, feel it and hear the 'psst' as it is opened before you drink the sweet, fizzy solution.

It's not enough then to think about knowledge as just information, we also need to think of how this information is portrayed (Merrill, 2002). This means that once we have established *what* we want student to learn through devising the key questions for the lesson, we then need to think carefully about *how* students should experience these ideas so that they can build an understanding of the concept in all of its dimensions – not just as words on a page (Figure 9.2). This means using diagrams, analogies, models and animations that we talk about and refer to. We also need to plan for students to encounter manifestations of the information through practical demonstrations, whole class investigations and videos. If, for example,

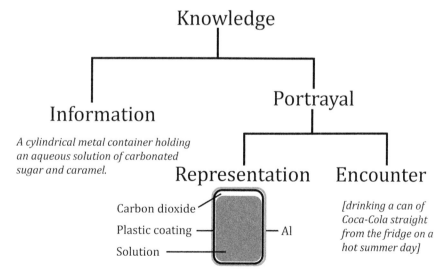

Figure 9.2 Knowledge is more than information and needs to be portrayed

we are teaching the word 'sonorous' students need to hear a bell being rung. These representations and encounters, though, are more than just bringing science to life. They also provide the objects through which students build their understandings from because they provide a common point of reference that can be talked about within the classroom and can form the basis for teacher questions to explore understanding.

Planning how information is portrayed

Having seen that knowledge is more than just information, let's return to the key questions from Chapter 8 and remind ourselves of what we want students to learn.

1. **Name three properties of metals and give a use for each**
   Good conductor of electricity – for making electrical wires
   High melting points (except mercury −38.8 °C) – for making jet engines
   Malleable – can be hammered into shape – for making jewellery
2. **Draw a labelled diagram of a metallic lattice and describe the differences between a metallic bond and an ionic bond**
   A metallic bond is the electrostatic attraction between the positive metal cations and the delocalised electrons (Figure 9.3). It is similar to an ionic bond because it is an electrostatic force of attraction; however, it is different to an ionic bond because the attraction is between positive ions and the sea of delocalised electrons and not between positive ions and negative ions.
3. **Explain why electrical wires are made from copper metal and not solid sodium chloride**

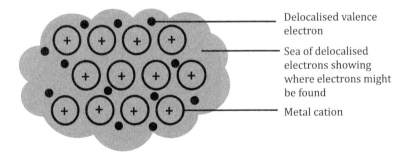

Delocalised valence electron

Sea of delocalised electrons showing where electrons might be found

Metal cation

Figure 9.3 A model of a metallic lattice

In copper metal the outer shells of adjacent atoms overlap. This means the valence electrons are not associated with any one atom and are delocalised, meaning they are free to drift throughout the lattice. When a potential difference is applied to the metal, the delocalised electrons can move throughout the lattice structure completing the circuit. Solid sodium chloride has no ions or electrons that are free to move and complete the circuit so cannot conduct electricity. Sodium chloride is also brittle and so cannot be shaped into wires.

We now need to plan carefully how this information is portrayed so that students relate these abstract words to specific situations that can be talked about and reasoned with.

Taking the key questions and expected answers, we consider possible misconceptions students might have – work by Driver *et al.* (1994) is a great place to start here. For ideas of how to explore chemistry misconceptions see Taber (2002). Now begins a process of matching, where we look at how we can best portray the knowledge needed to answer each key question (Table 9.1). This may well involve practical work such as a teacher demonstration, a whole class confirmatory experiment, investigation, observation activity or time to practice a specific scientific technique (see Holman, 2017). Finally, we consider how students are going to use this knowledge or skill to consolidate, extend and reveal their thinking.

The form: structuring learning

We have so far defined what we want students to learn and considered how this information will be portrayed and used in the lesson. Let's now think about how this knowledge can be taught, using what we know about the science of learning. This is what I am calling the 'form' of lesson planning and refers to the shape or organisation of the lesson to ensure that planning is informed not only by what we want students to know and do but also by what research tells us about how and why students learn in the first place.

The form comprises six phases (Figure 9.4). You may well adapt this model or even reject it as you become more confident in planning. I hope, though, that it provides a useful starting point to begin thinking and talking about the planning process.

Table 9.1 Matching the 'what' and the 'how'

Key question	WHAT? Expected answer (misconceptions)	HOW? Portrayed	Used
1. Name three properties of metals and give a use for each.	Good conductor of electricity. Making electrical wires. High melting points (except mercury –38.8 °C). Making jet engines. Malleable – can be hammered into shape. Making jewellery. (All metals are magnetic; metals exist only as solids.)	Students are given a piece of copper metal to explore. They are not told what the substance is but given melting point data. A simple circuit is provided with a bulb for students to test electrical conductivity. A magnet is also provided to address the misconception that all metals are magnetic.	Students must suggest uses for the 'unknown' substance, based on its physical properties.
2. Draw a labelled diagram of a metallic lattice and describe the differences between a metallic bond and an ionic bond.	See diagram in Figure 9.3. A metallic bond is the electrostatic attraction between the positive metal cations and the delocalised electrons. It is similar to an ionic bond because it is an electrostatic force of attraction; however, it is different to an ionic bond because the attraction is between positive ions and the sea of delocalised electrons and not between positive ions and negative ions. (Metals don't have bonding as they are elements; atoms lose electrons to get full shells.)	Teacher explanation. A metallic lattice is drawn on a whiteboard imagining we are zooming into the piece of copper students were exploring above. Bubble raft model is demonstrated in the classroom (https://www.rigb.org/our-history/bragg-film-archive/bubble-rafts/experiments-with-the-bubble-model) to represent the lattice structure with non-directional metallic bonding. Image shown to recap ionic lattice structure.	Students copy down the diagram and use this to label a picture showing the bubble raft model. Students complete a Venn diagram in pairs comparing bonding and structure of copper to sodium chloride. Statements are provided to scaffold learning.
3. Explain why electrical wires are made from copper metal and not solid sodium chloride.	In copper metal the outer shells of adjacent atoms overlap. This means the valence electrons are not associated with any one atom and are delocalised, meaning they are free to drift throughout the lattice. When a potential difference is applied to the metal, the delocalised electrons can move throughout the lattice structure completing the circuit. Solid sodium chloride has no ions or electrons that are free to move and complete the circuit so cannot conduct electricity. Sodium chloride is also brittle and so cannot be shaped into wires.	Teacher demonstration showing that an electrical cable is made from copper by cutting through an unplugged cable. Samples of copper metal and sodium chloride.	Students plan out and write an answer to key question 3.

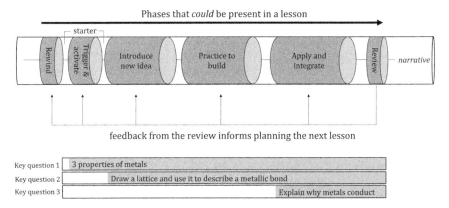

Figure 9.4 Phases of instruction that could be present in a science lesson or series of lessons

The six phases of instruction

The six phases of instruction are outlined below but we will take a closer look at each phase in more detail in Chapters 10–14. Phases indicated with asterisks are likely to be present in all lessons when you start out teaching and it's perfectly possible that you might choose to repeat certain phases within the same lesson. The timings are for guidance only as these will change depending on the activities that you choose to use.

1. ***Rewind** (5 minutes)
   The rewind provides an opportunity for students to retrieve some key ideas from previous lessons. The act of retrieval is important for memory but also should allow students to experience success. This time in the lesson is also an opportunity for the teacher to establish behavioural and academic expectations, deal with any issues such as no pens and take the register.
2. **Trigger interest and activate prior knowledge** (5–10 minutes)
   The lesson starts by triggering students' interest about the scientific ideas they are about to learn. This 'trigger' should also activate and reveal students' prior knowledge so that the teacher can build on the collective wisdom of the class. This part of the lesson then is not a test, it's a memory jog.
3. ***Introduce new idea**
   As we saw in Chapter 4, scientific ideas are often abstract, complex and counterintuitive. This is the time when students are introduced to the new scientific idea(s) with real care using a range of representations, encounters and clear explanations.

4. ***Practice to build understanding**
   Students need time to practice using the ideas once they have been introduced. This is often the aspect that gets overlooked, but sustained practice is important for students if they are going to explore ideas in sufficient depth and commit them to memory. It also creates time to help those students who are finding it hard whilst the rest of the class can get on. As we will see in Chapter 13, this practice looks different depending on whether students are practicing using conceptual or procedural knowledge.

5. **Apply and integrate**
   Now comes the time when we want to find out if students are mimicking or have really understood the ideas. We also want to give students the opportunity to use, defend, adapt and extend the scientific ideas to see their value and deepen their understanding. This probably will involve a task that requires ideas from the practice phase to be applied to an unfamiliar example and may involve drawing on knowledge from other powerful ideas. If students succeed, then this is quite good evidence that they are learning, if they struggle there is an opportunity for feedback.

6. ***Review and reflect** (5–10 minutes)
   Students see if they can answer the three key questions that were used to plan the lesson. Their answers are then used as the starting point for the next lesson. This doesn't want to feel high-stakes, but it does want to serve as a reminder for students to take responsibility for their learning by providing a time to reflect on what they have and haven't learnt. An alternative to using key questions here is to return to the question or demonstration used to trigger students' interest at the start of the lesson to see if they can now explain the idea.

Transitions and flow

Looking at Figure 9.4, there are gaps between each phase called transitions. When you are planning your lessons you will want to think about how to transition from one phase to another. For example, this may involve students moving from their seats to the middle desk to see a teacher demonstration and then back again. These transitions want to be as short and smooth as possible so no time is wasted. Establishing clear routines, such as always carrying out the demonstration in the same location and having the same system for students coming forward will help. Starting all instructions with 'In a moment' will prevent students from moving before you are ready.

Whilst transitions are inevitable, it's important that the lesson doesn't feel fragmented, but rather ideas flow and build from one phase to another just like a piece of music might. This can be achieved in two ways.

The first is to create a narrative or purpose that runs throughout the lesson and so threads it together. You can create this narrative by posing an overarching question or goal at the start. An example could be that 'today we going to find out why water is sticky' or 'why are electrical wires made from copper and not salt?' The question doesn't want to be overly complex, but it does want to generate a sense of curiosity and that makes it clear what area of science we are learning about today.

The second way is to ensure that key questions aren't introduced in one phase and ignored in the next. Instead, think of key questions being constantly built upon and connected together as a lesson progresses. Students are then not discarding the learning from one phase as they move to the next, they are adding to and refining it. In this way, learning deepens over time because knowledge learnt in one phase is built upon in the next, allowing time for consolidation of understanding to take place. In the lesson planned in this chapter, the lattice model was used repeatedly, but in more sophisticated ways throughout the lesson, allowing students to become confident in using and thinking about this model.

How this lesson looks on a plan for a 60-minute lesson

Just as not all music will contain the same sections, the phases present in Table 9.2 do not need to be in every science lesson. Indeed, it's possible and sometimes desirable for some lessons to focus on specific phases only. If you're teaching balancing equations, it would make perfect sense for students to spend much of the lesson practicing a range of examples. Indeed, you may prefer to think about the different phases as running not within a lesson, but across a unit of work that spans multiple lessons. Feel confident to adapt this planning framework to work for you, your classes and the content you are teaching and never see it as a list of things that must be present in every lesson.

Table 9.2 The overall plan for the lesson

Phase	Time (min)	What students are doing	What the teacher is doing	Key question
Rewind	5	Answering questions from the board recapping previous learning on atomic structure and ionic bonding	Using praise and sanctions to settle class Setting expectations for the lesson	Recap atomic structure and ionic bonding
Transition*		Group pairs and distribute equipment in trays		
Trigger interest and activate prior knowledge	10	Working in pairs to explore properties of an unknown substance (Cu) and suggest uses from physical properties	Managing pair focus and listening to discussions Review answers and clarify uses. Share that in the United Kingdom each person has about 175 kg of Cu associated with them, e.g. wires, cables – not enough Cu deposits in the world for all to have this	KQ1
Transition		Set-up bubble raft model		
Introduce	10	Listening to teacher explanation and answering questions Drawing lattice model into books and label diagram of bubble model	Explanation of metal structure and bonding using whiteboard Demonstrate bubble raft model Circulate to look at diagrams Ask students to review task by labelling latice on board	KQ1 KQ2
Transition				
Practice to build understanding	15	Work in pairs to complete Venn diagram to compare structure and bonding of Cu and NaCl (some ideas are provided to scaffold learning towards key question 3)	Share model of NaCl Circulate, support and listen to discussions Reveal answers	KQ1 KQ2

Table 9.2 The overall plan for the lesson (Continued)

Phase	Time (min)	What students are doing	What the teacher is doing	Key question
Transition		Hand out samples of Cu and NaCl		
Apply and integrate	15	Inspect Cu and NaCl and discuss answer	Pose key question 3 by cutting open (an unplugged) electrical cable to reveal the copper metal inside	KQ1
				KQ2
		Plan out and then write answer		**KQ3**
		Self-assess	Share success criteria	
Transition		Students pack away		
Reflect and review	5	Answer three key questions on scrap of paper	Collect in answers and decide starting point for next lesson	KQ1
				KQ2
				KQ3

* Transition time has been included in the time allocated to each phase. You may want to allocate transition time separately in your plan.
Bold in the table signifies when the key question is first introduced.

Assessment and rounding off tasks

We haven't explicitly looked at how tasks should be reviewed and how assessment for learning takes place in a science lesson. We will look specifically at feedback and responsive teaching in Chapter 15. However, it's worth mentioning now that whilst there are undoubtedly many tools we can use to try to see what learning has taken place, such as mini-whiteboards and voting cards, a lot of information on how students are doing comes from the tasks themselves. As you walk around the classroom during the rewind phase, you will get to see for yourself how much students are remembering. By listening to discussions during the activate phase, you will hear what students know already. That said, gathering this type of feedback when you first start teaching is tricky and so building in formal checks for understanding, such as those done using mini-white boards, can be a helpful way to get feedback. Where possible, look to review tasks properly, for example, by asking students to make corrections to their answers or peer assess work. This sends the message that tasks are valuable and that students must take responsibility to complete them. This will also create the opportunity to celebrate student work and build the sense that the class is learning together.

Summary

In this chapter, we have considered knowledge as comprising two aspects: information and portrayal. When planning science lessons we need to match the abstract ideas we want students to understand to specific representations and encounters in the classroom. This could involve using models, diagrams, explanations, practical work and a variety of other examples that students can use to reason with. In this way, students develop a meaningful understanding of the concepts they are learning about. By then thinking about the form of lesson planning, it is possible to use ideas from the science of learning to structure learning into one of six phases, each with a specific purpose that draws on what we know about motivation and cognition. In Chapters 10–14, we will look at each phase in a little more detail by drawing on examples and non-examples to illustrate what might work and why when teaching different powerful ideas. It is through understanding the rationale for each phase that you will be able to take this model of how to plan a science lesson and adapt it to match the needs of your students and the content you are teaching.

Bibliography

Driver, R., Squires, A., Rushworth, P. and Wood-Robinson, V. 1994. *Making sense of secondary science: Research into children's ideas.* London: Routledge.

Holman, J. 2017. *Good practical science.* London: Gatsby Charitable Foundation. Available at: www.gatsby.org.uk/uploads/education/reports/pdf /good-practical-science-report.pdf. [Accessed: 20 December 2019.]

Lederman, N. G., Antink, A. and Bartos, S. 2014. Nature of science, scientific inquiry, and socio-scientific issues arising from genetics: A pathway to developing a scientifically literate citizenry. *Science & Education*, 23(2), pp. 285–302.

Merrill, M. D. 2002. First principles of instruction. *Educational Technology Research and Development*, 50(3), pp. 43–59.

Taber, K. 2002. *Chemical misconceptions: Prevention, diagnosis and cure* (Vol. 1). London: Royal Society of Chemistry.

Section 4

Planning and teaching the phases of instruction

CHAPTER 10

Rewind and success for all
Retrieval practice

Powerful idea focus

- *Chemical reactions only occur if they increase the disorder of the Universe.*

Pedagogy focus

- Using retrieval practice to help memory.

Aims

The rewind is an opportunity for:

- students to practice retrieving ideas from previous lessons which helps memory,
- students to experience a quick sense of success at the start of the lesson and
- the teacher to deal with whatever needs to be done before the lesson begins such as distributing equipment, taking the register or dealing with late-comers.

Duration

- 5 minutes including feedback.

> ## ▌ Link to the theory
>
> - The rewind phase draws on two essential ideas. The first is the need to retrieve ideas from memory to prevent knowledge from being forgotten – that is, the concept of retrieval practice – and links to ideas discussed in Chapter 5 regarding learning and long-term memory. The second reason links to motivation, as answering questions correctly can foster feelings of competence among students, which links to what we learnt in Chapter 6.

Why is it that most chemical reactions mess things up? They destroy buildings, they corrode cars and they age us. There are at least three possible reasons why you might not be able to answer this question. Perhaps you were never told the answer and had no inclination to find it out for yourself. Or, if you were told, you never learnt or encoded the information in long-term memory, either because you weren't interested or because you couldn't make sense of it. Either way you will have, perhaps unfairly, been seen to have forgotten it! Finally, you may have learnt why chemical reactions happen, but just now you can't remember, the memory has slipped away.

But don't worry if you have forgotten, you are not alone. Forgetting is totally normal and humans are especially good at it. Professor Robert Kraft from Otterbein University suggests that humans don't approach the world with the aim of remembering it, rather we approach the world with a desire to understand it (Kraft, 2017). Indeed, forgetting seems to be the default position for most information we meet in our day-to-day lives. So, whilst it's likely you can remember what you had for breakfast this morning, you're unlikely to remember what you had for breakfast 23 days ago, unless you are someone who likes a routine. Forgetting this type of trivial detail is beneficial because it prevents the 'system', that is, our brains, from becoming overwhelmed with lots of superfluous information that gets in the way of remembering, or retrieving, the important things – the things that mean something to us.

Unfortunately, we do forget things that we need to remember. Every evening in the British Library there are always a few readers trying to remember which locker (of the 1,000 or so) they put their bags in earlier that day. I regularly experienced this problem whilst writing this book, but got around it by choosing a locker number that meant something to me (for security reasons I shall not be disclosing this number). So, it was the memory associated with the locker number which helped me to remember, rather than the number itself. This is called a retrieval cue and is how we remember lots of things and explains why you can remember what you had for break-fast the day something significant happened. Annoyingly though, this locker number wasn't always available and so then I had to take a locker with a number that was meaningless.

To help me remember this meaningless locker I would practice remembering the locker number at various points throughout the day, which for the most part worked. Unbeknown to be, I was using a strategy that has been known by psychologists for over 100 years, where simply the act of trying to retrieve information from long-term memory helps us to remember it. This act of remembering is called retrieval practice, or sometimes known as the testing effect (Roediger and Butler, 2011). Before we look at how this works in the classroom, let's first spend a bit of time thinking about what it is that we don't want our students to forget about the powerful idea of this chapter – that is, *chemical reactions only occur if they increase the disorder of the Universe.*

Why chemical reactions happen

To understand why chemical reactions mess things up, we need to under-stand why chemical reactions happen in the first place. I am not talking here about the rate of reaction – that is, how quickly reactants are turned into products; I am talking about whether the change can take place at all once we've supplied the initial energy to get the reaction started. Chemists refer to these as feasible reactions, which is just another way of saying that products are favoured over reactants. Can we, for example, take carbon and react it with hydrogen at room temperature to make benzene? Just because we can write an equation for this reaction doesn't mean it can happen.

$$6C(s) + 3H_2(g) \rightarrow C_6H_6(l)$$

$$\Delta H = +49 \text{ kJ mol}^{-1}$$

To answer this question, we're going to start off by thinking about why chemical reactions happen with a little thought experiment. Imagine you have a brand-new pack of cards. Now take the pack and shuffle it. Something important has just happened. In that very moment you have increased the disorder of the Universe, because you have made it much harder to predict where any one card in the pack will be. Shuffle again and the disorder will increase further; there is no way, or at least it's incredibly unlikely, that the pack will randomly organise itself. What you have just witnessed is the nat-ural tendency of our Universe to increase in disorder, simply because dis-order is more likely: cards get mixed up, perfume spreads out and sandcastles get washed away. Disorder always wins.

Although entropy can be quite a frightening word for students, it can be understood relatively easily using the analogy of disorder. A tidy room has a lower entropy than an untidy room and solid iodine has a lower entropy than iodine as a gas. Over time students will come to appreciate the limitations of this analogy, in that entropy considers both matter *and* energy; nevertheless, it's a useful starting point to build understanding from and provides 'an early flavour of what entropy means' (Haglund, 2017, p. 205). Sticking with this analogy, according to the second law of thermodynamics, chemical reactions can only happen if there is an increase in the disorder (entropy) of the Universe. Crudely put, this means that after a chemical reaction there must be less order/more disorder somewhere else in the Universe – we can use this idea to figure out if a reaction is feasible or not under a specific set of conditions.

This all seems slightly mysterious, but it's easier to understand entropy by thinking about the Universe in two separate parts – the system that is made up of the chemicals being studied and the surroundings, such as apparatus and air in the laboratory, which makes up everything else. We assume that matter cannot move from the system to the surroundings, only energy can (Figure 10.1).

It's always easier working with tangible examples in science, so let's take the reaction of magnesium with oxygen. We know this reaction is feasible because we've all been enthralled by the bright white light that is produced when magnesium burns in air, but let's think about why this reaction can happen in the first place.

A quick look at the equation (Figure 10.2) would suggest that the reaction shouldn't occur because the disorder of the chemicals decreases as a solid

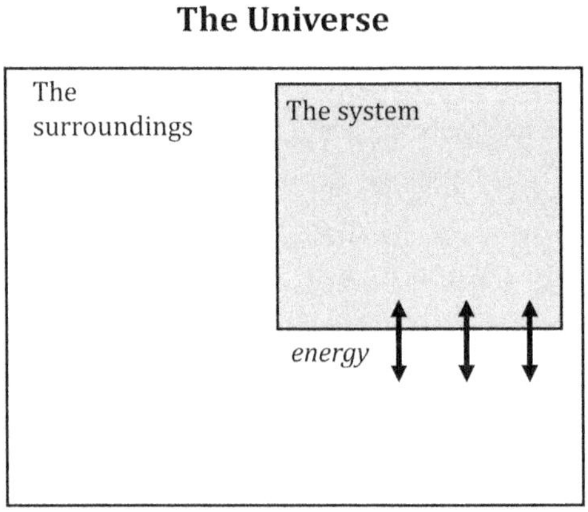

The Universe

Figure 10.1 Thinking about the Universe in terms of a system and its surroundings

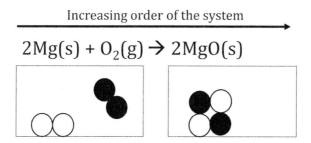

Figure 10.2 When magnesium reacts with oxygen there is a decrease in the entropy (disorder) of the system

and a gas react to form a solid. So far though we have only considered the chemicals – the system – but we also need to consider what happens to the surroundings.

When magnesium metal reacts with oxygen a lot of energy is released because the products are much more stable than the reactants. This means that energy is transferred from the chemicals in the system to the particles in the surroundings (Figure 10.3), which then move around more. The reaction is described as exothermic as energy transfers out of the system as it spreads out.

So, even though the disorder of the system decreased when solid magnesium and gaseous oxygen combined to form solid magnesium oxide, this was compensated for by the increase in disorder of the surroundings.

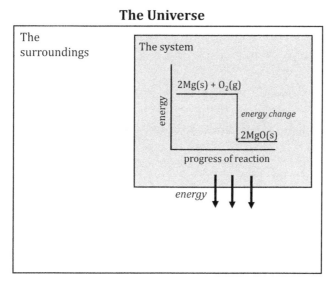

Figure 10.3 When magnesium reacts with oxygen, energy is transferred out of the system to the surroundings which increase in entropy

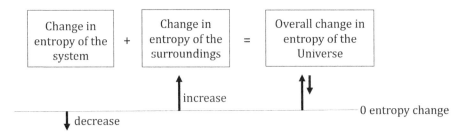

Figure 10.4 The entropy changes when magnesium reacts with oxygen

Overall the disorder of the Universe increased under those conditions and so this explains why the reaction can occur according to the second law of thermodynamics. It's almost as if the increase in entropy of the surroundings pays for the decrease in entropy of the system (Figure 10.4).

So what about endothermic reactions? How is it possible that reactions that cool the surroundings down, such as those used in ice packs or Sherbet Dip Dabs, are feasible? The thermal decomposition of copper carbonate is a lovely example of a feasible endothermic reaction that only occurs when it is furiously heated with a Bunsen burner. The fact we have to continuously heat copper carbonate for the reaction to occur means that the entropy of the surroundings must decrease, as energy is transferred into the system. However, if we look at the equation for this reaction, we can see that it produces a gas (Figure 10.5).

Gases are more disordered than liquids or solids, meaning that the entropy of the system increases. This increase in the disorder of the system is enough to compensate for the decrease in disorder of the surroundings, meaning that overall, there is an increase in the disorder of the Universe so the reaction can happen.

So, to summarise, chemical reactions (and physical changes) mess things up because they can only happen if they increase the disorder of the Universe – this depends on both enthalpy and entropy changes. This means

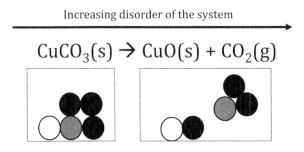

Increasing disorder of the system

$$CuCO_3(s) \rightarrow CuO(s) + CO_2(g)$$

Figure 10.5 Thermal decomposition of copper carbonate produces a gas

that if a reaction is endothermic and the products are more ordered than the reactants, then the reaction is not feasible at any temperature and would need to be coupled with another reaction that is feasible. For example, organisms must consume, and react together through respiration, vast amounts of food and oxygen to maintain order. Whilst there is more to this story than has been described here, in terms of free energy and how entropy changes are calculated using temperature, the fundamental idea holds: 'a reaction will go if the total entropy of the system and its surroundings increases' (Gillespie, 1997, p. 863). If that's not a powerful idea I don't know what is!

Going deeper

We could stop there, but we're just getting to the interesting bit so let's keep going. On the surface it's a little confusing that both reactions above needed to be heated, yet the reaction of magnesium was exothermic, whereas the decomposition of copper carbonate was endothermic, taking energy in. This conundrum, or conflict, gets us thinking about an important conceptual distinction here between the purpose of heating these two compounds.

Magnesium was 'dipped' into the Bunsen flame to get the reaction going, just like a match is used to light a candle. Once the chemical reaction starts, the magnesium can be removed from the flame and the reaction with oxygen will continue, just like a ball will continue to roll down a hill after an initial push. This is why we only need one match per candle – the match is simply needed to overcome what is called the activation energy. This is the minimum energy needed for a reaction to occur and involves the breaking of bonds in reacting particles. All reactions have an activation energy, just these differ in size. However, if copper carbonate is removed from the flame the reaction will stop. Magnesium, then, was heated to get the reaction going, whereas copper carbonate was heated to start the reaction *and* then to sustain the reaction temperature.

The 'art' of remembering

Now that we have seen what is important about this powerful idea, let's consider how we can help students to remember all of these ideas, or at least forget less of them – there are at least two things we can do to help. The first is to help students learn these ideas in a meaningful way in the first place. We will look at how to do this in subsequent chapters but in essence this involves introducing new ideas by connecting them to what students already know. Our focus here is to consider how we can help students to remember these ideas once they have been committed to long-term memory. For this, we are going to need the help of retrieval practice that we met earlier when thinking about lockers in the British Library.

Retrieval practice

Retrieval practice, often called the testing effect, uses the process of testing as a learning tool as opposed to an assessment tool (see Roediger and Butler, 2011). For the most part, assessments are used in schools to find out what students have and have not learnt. However, various studies (see, e.g. McDaniel *et al.*, 2011; Roediger and Karpicke, 2006) have shown that testing – that is asking students to recall information from memory – helps students to remember ideas for longer. Importantly, this is more effective than simply re-studying the material. So, rather than asking you to re-read this chapter, I am going to quiz you, by asking you to predict (i) whether the diagram in Figure 10.6 shows an exothermic or endothermic reaction and (ii) whether this reaction is feasible or not under these conditions.

Effective quizzing requires students to take part in effortful processing when they retrieve information and that this retrieval processes is spaced out over increasing periods of time, often called spaced review (Pomerance *et al.*, 2016). So students should do quizzes shortly after learning the material and over progressively longer and longer periods of time after there has been time to forget. The term testing effect is in some ways not ideal, as it conjures up ideas of high-stakes assessment. Instead, it's best to think of this as the quizzing effect. Some of the key principles of how to use retrieval practice are described below:

i.   Questions should involve some level of processing.
ii. Correct answers should be provided afterwards to allow feedback – so, to answer the question from Figure 10.6 – the reaction is endothermic

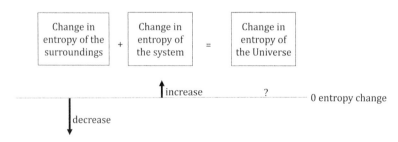

Figure 10.6 Predict whether this reaction is feasible and whether it is exothermic or endothermic

because the entropy of the surroundings decreased and it is not feasible (possible) under these conditions because the decrease in entropy of the surroundings is greater than the increase in entropy of the system. This reaction would be feasible at higher temperatures, but to understand why we would need to explore how entropy is calculated. This is beyond the scope of this chapter but is well worth an exploration.

iii. Quizzing should take place over progressively longer and longer periods of time.

Whilst you can implement retrieval practice in any part of the lesson, and students can make use of this learning principle when making revision cards or completing booklets, we are going to consider what this looks like in the context of the rewind phase that takes place at the start of the lesson.

An example

Good morning everyone, we are going to start with a quick review of some important ideas we are going to need today. Working on your own, in silence, you have 5 minutes to try and remember these ideas as you answer the questions.

Students work individually to answer the questions in silence (Figure 10.7). Students are encouraged to get started by praising those who have already begun and swiftly prompting those who haven't. Once 80% or more of the

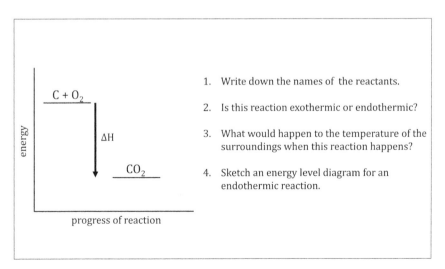

Figure 10.7 An example of a rewind activity

class are working, the teacher walks around the class looking at answers briefly to inform how they will review the task. After 5 minutes the class is stopped, even if everyone hasn't quite finished. Answers are reviewed quickly using no-hands-up questioning, these are celebrated with praise and correct answers are written onto the board. Some teacher questions are posed to explore reasoning, for example asking individuals to explain their answer to question two. Students correct *and* extend their answers. If any misconceptions emerge, these are noted down but are dealt with at a later stage in the lesson or in subsequent lessons once there has been a chance to plan an appropriate response. You may want to select one or two students to come to the board to draw out their answer to question four as this will provide an opportunity for feedback.

Design principles

The example above assumes that students have already learnt about exothermic and endothermic reactions. It uses a number of design principles that we will now explore in more detail so that you can transfer these ideas to any lesson.

Make sure the knowledge being retrieved is fundamental

There are lots of questions you can ask students about energy changes that take place during chemical reactions, but focus on the concepts that are most important for understanding the scientific ideas. In this example, it is all about making sure students can interpret an energy level diagram and can understand that exothermic reactions transfer energy to the surroundings and so warm them up. It also drew on other areas of relevant knowledge such as combustion and formulae and so retrieved key knowledge from different powerful ideas.

Plan for progression in mind

Whatever task you create, you want there to be progression built into it. This serves two purposes. First, it helps learning because each idea builds from the next in a way that scaffolds understanding and makes sense. The second purpose is to make sure everyone in the class has the opportunity to feel successful. In the example above, the first question orientates student thinking and is relatively easy. The questions that follow get progressively harder as they become more abstracted away from the diagram on the left but they remain connected by a common concept – in this case energy level diagrams.

Avoid unnecessary clutter

The task should all be on one slide, or it could be printed onto a piece of paper for students to complete. The problem with printing is that it increases teacher workload in terms of photocopying and sticking these sheets into books takes valuable lesson time. A better approach is to get students used to doing the rewind from the board. If you teach in different rooms, it is helpful to write the rewind on the teacher whiteboard so you don't have to wait to login.

Leave room for the answers

Underneath each question there is space to write the answer when you review the task. Having the answer underneath the question allows students to clearly see which answer refers to each question. Try to avoid showing all answers at once as this can be a little overwhelming and means students tend to just copy answers as opposed to correcting and thinking about each answer separately. It also removes the opportunity to celebrate students' answers.

Although we didn't consider entropy explicitly here in this example, this highly exothermic reaction causes an increase in the entropy of the surroundings, meaning that overall this reaction increases the entropy of the Universe and so does occur, which is good news for those of us who enjoy charcoal BBQs.

A non-example

We learn concepts through examples, but we also use non-examples to understand what a concept is not. For example, to understand what a dog is, it is helpful to know why a cat is not a dog. In this case, the cat is a non-example. I am therefore going to use some non-examples in the following chapters to help illustrate what each phase of the lesson is trying to do. By doing this, I am not identifying 'bad activities'; rather, I am trying to help you to understand what this phase of the lesson doesn't look like. Below is a non-example of an activity that could be used during the rewind (Figure 10.8). Before we discuss its differences to the example above, consider what you think are its limitations.

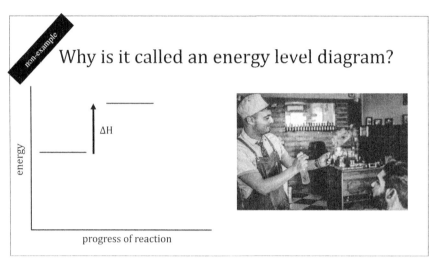

Figure 10.8 A non-example of a rewind activity

Source: Photo by Nick Karvounis on Unsplash.

Why is this a non-example of a rewind?

1. The question is very open and is not asking students to retrieve fundamental knowledge.
2. There is only one question and it's quite hard. There is no scaffolding built into this task to support students to be successful.
3. This task would need to involve discussion to explore ideas.
4. The image on the right is showing an exothermic reaction but this is not relevant to the question which is about energy level diagrams and is showing an endothermic reaction. A better image might be a block of flats, or diving boards that clearly show different floors as this supports the right thinking here.
5. No substances are on the energy level diagram making it very abstract.

Challenges

Perhaps the most common mistake of the rewind phase is that it lasts too long. The symptom is pace but the cause is most often a planning one. Either the task is asking students to retrieve knowledge they don't have, which means too long is spent doing and reviewing the task, or there is not sufficient clarity over what students need to do and so they hold back and don't commit to answering any question in fear of getting it wrong. In the beginning, try to focus on establishing a routine for how the rewind is carried out. Then stick with this routine so that time is spent thinking about the science and not how the activity is being done. Modelling the first answer is helpful if students are getting stuck on what to do, and, if in doubt, select easier questions in the beginning as the aim is not to confuse students here, there is plenty of time for that later!

Taking it further

As you develop confidence with the rewind, you may want to mix it up a little by using:

1. **Key word spread:** lots of key words on the board and students have to write a paragraph using as many key words as possible to explain a specific question. This can be difficult to review and provide feedback on, so go carefully.
2. **What's wrong and why?** There is an incorrect diagram on the board that students must redraw and label to make it correct, for example, an incorrect energy level diagram. It's really important that students have the opportunity to see the correct answer afterwards.
3. **What was the question?** There are answers on the board and students must write out what the question was.
4. **Not starting the lesson with a rewind.** Students could go straight into the practical or other task. This is helpful if time is limited.

Summary

No matter how well something is taught, it is inevitable that ideas will be forgotten. The rewind is a time in the lesson where you help students to remember by asking them to retrieve *the most important* ideas. Not only does this process of retrieval help students to remember, it also creates a sense of progress at the start of the lesson which is an important source of motivation. There is also the pragmatic advantage that it creates time at the start of the lesson to sort out any admin tasks and get ready for the next phase of the lesson, the time where we trigger interest and activate prior knowledge.

Bibliography

Gillespie, R. J. 1997. The great ideas of chemistry. *Journal of Chemical Education*, 74(7), pp. 862–865.

Haglund, J. 2017. Good use of a 'bad' metaphor. *Science & Education*, 26(3–4), pp. 205–214.

Kraft, R. N. 2017. *Why we forget. The benefits of not remembering*. Available at: https://www.psychologytoday.com/gb/blog/defining-memories/201706/why-we-forget. [Accessed: 20 December 2019.]

McDaniel, M. A., Agarwal, P. K., Huelser, B. J., McDermott, K. B. and Roediger, H. L. 2011. Test-enhanced learning in a middle school science classroom: The effects of quiz frequency and placement. *Journal of Educational Psychology*, 103(2), pp. 399–414.

Pomerance, L., Greenberg, J. and Walsh, K. 2016. *Learning about learning: What every new teacher needs to know*. Washington, DC: National Council on Teacher Quality. Available at: https://www.nctq.org/dmsView/Learning_About_Learning_Report. [Accessed: 5 January 2020.]

Roediger, H. L. and Butler, A. C. 2011. The critical role of retrieval practice in long-term retention. *Trends in Cognitive Sciences*, 15(1), pp. 20–27.

Roediger, H. L. and Karpicke, J. D. 2006. Test-enhanced learning: Taking memory tests improves long-term retention. *Psychological Science*, 17(3), pp. 249–255.

Trigger interest and activate prior knowledge

Powerful idea focus

- *Every particle in our Universe attracts every other particle with a gravitational force.*

Pedagogy focus

- Triggering interest and activating and revealing prior knowledge.

Aims

The activate phase provides an opportunity to:

- develop interest in the subject matter by identifying the intrinsically interesting bits,
- activate relevant prior knowledge and mental models needed for the lesson so students connect new ideas to what they already know and
- reveal and explore students' prior knowledge and misconceptions.

Duration

- 5–10 minutes.

Link to theory

- This phase of the lesson draws on ideas of student interest and motivation that we met in Chapter 6, misconceptions from Chapter 4 as well as recognising the importance of building upon prior knowledge that links to what we learnt in Chapter 5 about how new knowledge is learnt.

In my left hand imagine I have a golf ball and, in my right, a ping-pong ball. Which ball will hit the ground first? I am going to give you a minute or so to write down your prediction and if you can, your explanation.

Imagine, I have now dropped both balls from the same height – they will appear to hit the ground at the same time. But how is this possible? How can two objects with different masses fall at the same rate? After all, heavy objects have a greater mass and therefore experience a greater pull towards Earth due to gravity. If you don't believe me, look at the calculations below where the weight of a golf ball is nearly 15 times greater than the weight of a ping-pong ball (Figure 11.1).

Now, if we were in the classroom we could ask two students to come to the front of the class. Each is given a straw and strict instruction to move either the ping-pong ball or the golf ball from one end of the table to the other without touching it. After some serious puffing, the golf ball barely moves, yet the ping-pong ball has reached the end of the table. You could repeat the experiment just to make sure that it's not a 'fluke' – a nice example of repeatability. We could get a different pair of students up to see if they can repeat the result – this is the idea of reproducibility. In both cases, I predict you will observe the same result – the ping-pong ball moves further. But why, and how is this related to the rate at which objects fall towards the Earth?

To answer this conundrum, you need to consider that the golf ball, with a larger mass, needs a greater force to get it moving in the first place – this is true whether the ball is rolling along a table or whether it is pulled towards

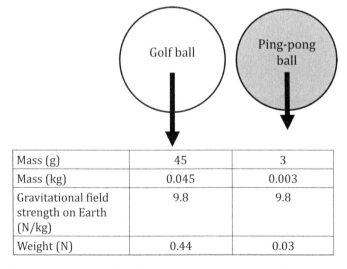

Mass (g)	45	3
Mass (kg)	0.045	0.003
Gravitational field strength on Earth (N/kg)	9.8	9.8
Weight (N)	0.44	0.03

Figure 11.1 The weight of an object is the force of gravity pulling it towards Earth

Earth after it is released from your hand. The ping-pong ball in contrast, with a smaller mass, only needs a smaller force to get it moving. This fits with our everyday experiences in that it is easier to push a car than a lorry to get it moving. The scene has now been set to explore Isaac Newton's powerful idea of gravitation, the idea that explains why apples fall, moons orbit and planets form.

Teacher demonstrations to focus attention

Demonstrations like the falling balls play an important role in triggering students' interest in science. Interest is important in the short term because it helps to focus students' attention on specific aspects of the scientific idea that we want them thinking about. In the demonstration above we focused in on the significance of mass to challenge the common sense, but wrong, idea that heavy objects fall faster than light ones.

We could have triggered interest by shouting 'bang!' at the start of the lesson, though – that would surprise students! Unless your lesson is on sound or responses to the environment, though, this surprise is getting students thinking about the wrong things. Good triggers, then, are moments that use the inherently interesting aspects of the subject to arouse interest. McCrory (2011) refers to such ideas as internal hooks. By triggering interest in this way, we encourage students to invest effort into thinking more deeply about the scientific ideas themselves (Harackiewicz *et al.*, 2016).

Whilst the ball-drop demonstration above seemed relatively simple, it is actually doing a number of clever things that you can apply to any teacher demonstration.

1. It had a clear goal.
2. It directly challenged a common misconception.
3. It involved familiar objects that aroused interest and drew on students' prior knowledge.
4. It was simple – and removed the need for distracting apparatus.
5. It provided a context through which to explain the scientific ideas and could be referred to throughout the lesson.

Interest in the short and long term

Teacher demonstrations like this trigger something that we call situational interest. Situational interest refers to those moments in the classroom where students' interest is generated by some specific object or stimulus. This might include the demonstration we started this chapter with or perhaps showing the video of David Scott, the Apollo 15 astronaut, dropping a falcon feather and a hammer on the Moon and seeing that both objects hit the ground at exactly the same time. These experiences tend to be highly interesting but generally are short lived. The other way in which interest is experienced is when students want to reengage with a particular object or topic over time (Hidi and Renninger, 2006). So, for some students, the subject matter itself of falling objects is already intrinsically interesting – they don't need any convincing.

The goal then of instruction is for situational interest to be sustained over time (Figure 11.2), across many different situations to eventually become an individual interest (Hidi and Renninger, 2006). This is important because we want students to not only know more science, but also to experience a love for this subject. This extends beyond fleeting moments of surprise but develops into something more permanent that changes students' beliefs and attitudes about the value of science itself. Perhaps culminating in a desire to study this subject at University and beyond, irrespective of whether it is assessed or not.

So, we may spark students' initial interest in gravity by blowing a ping-pong ball and a golf ball along a table, but then will need to maintain this interest through well-planned lessons that allow students to understand and make sense of the scientific ideas beyond the interesting 'hook'. We need then to create opportunities for students to see the power of scientific ideas – that is, to see how scientific ideas such as gravity allow them to explain a range of phenomena situated over a range of contexts. In this way, students

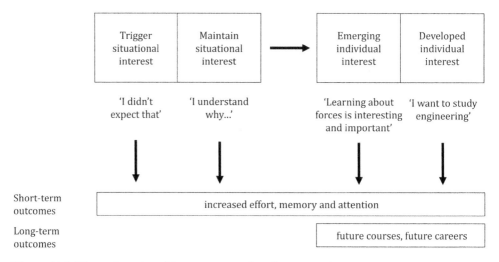

Figure 11.2 How situational interest may develop over time

will start to see gravity and physics as not only interesting but as relevant and important to their everyday lives.

With this in mind, let's explore further about what is powerful about the scientific idea that *every particle in our Universe attracts every other particle with a gravitational force*. We will then look at ways to trigger interest and activate prior knowledge in the classroom.

Why gravity matters

The story of our solar system began around five billion years ago with a bang, albeit a silent one! This 'bang', not to be confused with the origin of the Universe, was a nearby star exploding that started a solar nebula, a swirling cloud of dust and gas. Gravity, the force of attraction between all objects with mass, then set about assembling all of this debris together, forming the planets we see today. At the heart of this giant nebula vast temperatures and pressures caused hydrogen nuclei to fuse to form a new star – our Sun.

The Sun, with its enormous mass, kept the rest of our solar system in place and still does today. Without this gravitational force the celestial bodies, including Earth, would simply have drifted off into the depths of space like a ball rolling along a frictionless table. So, why doesn't Earth get pulled into the Sun; after all, Earth has a much smaller mass?

To understand what is happening here it's best to return to Chapter 4, where we learnt about Newton's first law. That is, a force is needed to change the direction or speed of an object. Or, in other words, motion in a straight line is the natural state. So, without gravity Earth would simply be moving through space in a straight line until it bumps into something. However, Earth is prevented from moving in a straight line because it is attracted towards the Sun due to gravity. This 'tug', which keeps Earth moving in a circular orbit, is called the centripetal force and acts towards the centre of circular motion. But because Earth is constantly moving through space in a sideways direction, thankfully this 'tug' from the Sun isn't sufficient for us to crash into it and burn up.

For around 100 million years, Earth orbited the Sun without a moon. Then, around 4.5 billion years ago, a massive astronomical body the size of Mars smashed into Earth, throwing out rocks and debris. Thanks again to gravity, this debris assembled together to form our Moon.[*] This collision also knocked Earth over somewhat, causing Earth to orbit the Sun whilst it rotates on a tilted axis of 23.5°, which it continues to do today.

---

[*] At least that's one hypothesis.

As well as giving us our Moon, this random collision also gave us our seasons. For half of the year, this tilt of 23.5° means that the northern hemisphere faces towards the Sun and so experiences more direct sunlight, meaning summer months are much warmer than winter ones. You can prove why tilt matters for yourself by shining a torch on a piece of paper either directly, or at an angle, and then feeling how warm the surface gets.

The Moon is not our only companion in the solar system, though. There are another seven planets, dwarf planets, asteroids, comets and meteoroids that all orbit our Sun (NASA, n.d.). The other three terrestrial planets, Mercury, Venus and Mars, are made from rocky material with solid surfaces. These planets were able to withstand the enormous temperatures near the Sun when our solar system was first formed. The remaining planets, called the Jovian planets, were formed from ice, liquid and gas. Jupiter and Saturn are gas giants, whereas Uranus and Neptune are ice giants. Orbiting the terrestrial planets is the asteroid belt and orbiting the Jovian planets is the icy Kuiper belt, the home of many dwarf planets and asteroids. At the edge of our solar system is the predicted Oort cloud – a vast spherical shell – like a thick bubble – containing icy bodies as large as mountains where the gravitational influence of our Sun ends.

Beyond the Oort cloud lies the rest of the Milky Way, comprising thousands of planetary systems that also orbit stars just like our own. Some of these exoplanets – planets beyond our solar system – have already been discovered. At the centre of the Milky Way lies a supermassive black hole that all stars in the Milky Way orbit, including our own. Beyond the Milky Way our nearest galaxy is Andromeda, perhaps one of just one hundred billion galaxies in the Universe, one of which may quite possibly contain life.

The power of gravity then, like all powerful ideas, comes from its explanatory clout, being able to explain seemingly unrelated phenomenon such as the origin of our solar system, why objects fall and why we have seasons. This means there are no shortage of ideas from which to draw from when looking for ways to arouse interest in lessons.

Ways to trigger situational interest

Taking some of these ideas, let's look at how we can use them to trigger interest at the start of a lesson by drawing on the intrinsically interesting aspects of the subject matter in ways that engender different emotions. Developing these 'hooks' in collaboration with colleagues can be a good way to develop pedagogy and subject knowledge (McCauley *et al.*, 2015).

Using amusement: show this video of graduate students trying to explain the seasons by incorrectly over emphasising the elliptical orbit of Earth around the Sun (https://www.learner.org/series/a-private-universe/). Seeing these Ivy League graduates struggling to explain this everyday phenomenon is quite funny, especially considering they were dressed in their academic regalia. It's also quite interesting how academically successful students have learnt science in a way that they can't then use.

Using surprise to address common sense misconceptions: show students the clip of Brian Cox dropping a bowling ball and a feather on Earth in the presence and absence of air particles (https://www.youtube.com/watch?v=E43-CfukEgs). This demonstrates nicely how all objects, regardless of their mass, fall at the same rate when there are no air particles to get in the way.

Using wonder: show students the Pale Blue Dot, a photograph taken by the Voyager 1 space probe from the edge of our solar system on Valentine's day, 1990 (https://solarsystem.nasa.gov/resources/536/voyager-1s-pale-blue-dot/). The probe was just beyond Neptune, some 6 billion kilometres away from Earth, when it was ordered to turn around and snap the photo. Earth, the pale blue dot, is hardly visible in this photograph but this only makes it more wonderful as we admire the sheer scale of what we are looking at.

Using relevance: draw on ideas that matter directly to students by sharing items from the news. For example, on Thursday 24 July 2019, a particularly sneaky asteroid named 2019 OK came within 40,400 miles of Earth travelling at 55,000 miles per hour. This asteroid was only detected earlier that day by an observatory in Brazil. The Center for Near Earth Object Studies (CNEOS, 2019) reports that if this asteroid were to have collided with Earth the devastation could have reached an area of 50 miles across.

Using fantasy: ask students to predict what would happen if the Moon didn't exist.

Activating and revealing prior knowledge in the classroom

Activities like this do more than just trigger students' interest, though, they, also activate prior knowledge, reminding students what they already know about an idea. By designing activities carefully, not only can we trigger interest and activate prior knowledge at an individual level, we also reveal the collective ideas of the class at the start of the lesson. This serves two important roles. Firstly, it reveals what students already know about a concept so we can adapt subsequent phases of the lesson if necessary. The second, and perhaps more important role, is it draws together prior ideas from across the class making the case that learning is a social process and that all ideas are valued, not just the teacher's. This has the effect of making sure that all students feel motivated to participate.

How then can we achieve this sense of learning as a collective and bring ideas together? The trick is to design the task so it encourages discussion and reveals what students are thinking about. In general, more open and unguided approaches are most effective so that you don't unwittingly exclude students from taking part in this thinking. So, once you have identified what's intrinsically interesting about the subject matter, it's time to think about how you are going to explore students' thinking using that question, demonstration or other hook. Below are some activities that can be adapted for any subject, but I am exemplifying them here in terms of gravity.

Write down everything you know about – asking questions with specific goals often cause students to focus on getting the answer as opposed to learning more generally. This is a particular problem at the start of a lesson where students might know different things about a concept or idea. One way to get around this problem is to ask questions that have a non-specific goal. This is called the goal-free effect and encourages the use of goal-free problems (see, e.g. Sweller *et al.*, 2019). So, instead of asking students to draw free body diagrams of falling objects, we could show them a force diagram and ask students to *write down all the information you get from this diagram* (Figure 11.3).

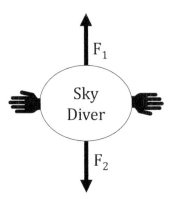

Figure 11.3 Write down all the information you get from this diagram; this is called the goal-free effect

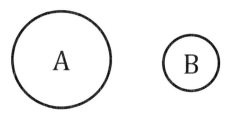

1. A pulls B with a greater force
2. B pulls A with a greater force
3. B and A pull each other with the same force
4. It's impossible to tell

Figure 11.4 By writing and discussing the answer to this question, students reveal their thinking about gravity; the answer is 3

Write-discuss-share – ask students to write down their answer or prediction to a challenging question (Figure 11.4). They then discuss their answer with a partner and add to their original answer, possibly refining or adding to it. Answers are then collected by the teacher through questioning. The activity ends when the ideas from the class have been put onto the whiteboard by the teacher. These ideas can be returned to at various points throughout the lesson. By using multiple-choice questions you can focus students' thinking on particularly important aspects of the concept. In the case of Figure 11.4, this allows us to see if students can apply what they know about gravity. This new context often causes problems, with some people incorrectly thinking that the question can't be answered as the mass of the two balls is not known. You can explore why this is not true using this PhET simulation: https://phet.colorado.edu/en/simulation /gravity-force-lab-basics.

Drawing – ask students to draw a diagram or annotate a picture to show what they think is happening. For example, students could draw a picture to show the forces acting on a ball that is travelling through the air. Many students will incorrectly add a 'push' force, whereas we know that only weight, and a small amount of air resistance is acting.

Sketching – ask students to sketch what they think a graph might look like. For example, students could work in pairs to sketch a graph that shows what is happening to the velocity of a skydiver at different points of their jump – you may want to give students the axes to scaffold learning.

▨ Design principles

When designing your activities to trigger interest and activate prior knowledge, the following design principles should help you to create tasks.

1. **Focus attention on intrinsically interesting aspects of the subject**
   Start with something physical, visual, or cognitive that grabs and focuses attention. This often involves addressing a common misconception, for example, that sky divers go up when the parachute is opened – they don't, they just appear to because they are being filmed by someone who hasn't opened their parachute yet. This creates a reason for wanting to learn.
2. **Ground the task in prior knowledge and make accessible for all**
   Use examples that connect to what students already know. This could involve linking to previous learning, or to everyday experiences.
3. **Foreshadow the lesson**
   This phase of the lesson should provide a snapshot of the extended lesson that follows. In this chapter, we used a demonstration to show that two balls with different masses fall at the same rate. This idea can then be unpacked throughout the lesson and returned to at the end in the review phase.
4. **Make thinking visible (and audible)**
   A hook serves at least three purposes. It should trigger students' interest, activate prior knowledge and reveal these ideas to the teacher and class. This means we need to think carefully about how tasks are designed so that they allow us to see or hear what students are thinking about.

▨ A non-example

Just as we did for Chapter 10, let's look at a non-example to see what might not work and why for this phase of the lesson. The non-example asks students to unscramble some key words to find out the topic for the lesson (Figure 11.5).

The problem with this activity is that it is not using what is intrinsically interesting about the subject matter to introduce the ideas. Rather, the interest is coming from being able to complete the task first. Whilst there is a stretch question, it's a question that you either know or you don't, and so it's encouraging students to guess as opposed to being an opportunity to use prior knowledge.

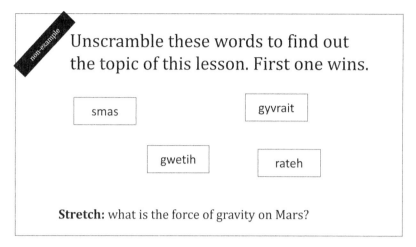

Figure 11.5 A non-example for triggering interest and activating prior knowledge in a lesson

Challenges

The beginning of a lesson can be a really motivating time for students and teachers when you get to look at some interesting aspects of science. Sometimes this means that this phase can take too long and asks students to think about too many ideas. When you're looking to trigger interest, keep the activities simple and focus thinking quickly on the interesting bits of the scientific idea by removing unnecessary noise. The golf ball and the ping-pong ball illustrate beautifully the power of simplicity. Whilst this phase of the lesson may reveal a number of scientific errors, this is generally not the place to resolve complex misconceptions, but rather a place to reveal them before teacher instruction begins. So, be confident to leave and then return to certain incorrect ideas once you've had a chance to introduce the new information.

Taking it further

1. Change the location of the hook in the lesson – it doesn't always need to be at the start.
2. Return to the hook at the end of the lesson to show students how their understanding has developed.
3. Get students to devise their own hooks to introduce a concept. This starts with a storyboard and could be videoed (McHugh and McCauley, 2017).

Summary

Hooks play a number of important roles in science teaching. They offer a bridge from everyday tangible examples to the abstract scientific ideas students are learning about. They help to activate and reveal prior knowledge. As we saw in Chapter 10, not only does this activation help memory, it also helps students to connect new ideas with what they already know and so encourages meaningful learning. Perhaps though, the most important role of hooks is to reveal to students the intrinsically interesting bits of science that increases situational interest, which in turn focuses attention and increases effort. Over time, this situational interest can be developed upon so that students see a value in the subject beyond these short-term moments of intrigue. By revealing and capturing what students are thinking about within this phase of the lesson, we gain information on where to start the next phase, the phase where new ideas are introduced.

Bibliography

Center for Near Earth Object Studies (CNEOS). 2019. *Largest asteroid to pass this close to Earth in a century.* Available at: https://cneos.jpl.nasa.gov /news/news203.html. [Accessed: 30 December 2019.]

Harackiewicz, J. M., Smith, J. L. and Priniski, S. J. 2016. Interest matters: The importance of promoting interest in education. *Policy Insights from the Behavioural and Brain Sciences*, 3(2), pp. 220–227.

Hidi, S. and Renninger, K. A. 2006. The four phase model of interest development. *Educational Psychologist*, 41, pp. 111–127.

McCauley, V., Davison, K. and Byrne, C. 2015. Collaborative lesson hook design in science teacher education: Advancing professional practice. *Irish Educational Studies*, 34(4), pp. 307–323.

McCrory, P. 2011. Developing interest in science through emotional engagement. In *ASE guide to primary science education*, Harlen, W. (ed.), 94–101. Hatfield, UK: Association for Science Education.

McHugh, M. and McCauley, V. 2017. Hooked on science. *Science in School*, 39, pp. 55–59.

NASA. n.d. *Solar system exploration.* Available at: https://solarsystem.nasa .gov/solar-system/our-solar-system/overview/. [Accessed: 15 December 2019.]

Sweller, J., van Merriënboer, J. J. and Paas, F. 2019. Cognitive architecture and instructional design: 20 years later. *Educational Psychology Review*, 31, pp. 261–292.

CHAPTER 12

Introducing new ideas
Explanations and models

Powerful idea focus

- *The total amount of energy in the Universe is always the same but can be transferred from one energy store to another during an event.*

Pedagogy focus

- Introducing new ideas using teacher explanations and models.

Aims

The introduce phase allows scientific ideas to be introduced so that:

- they build from, and connect to, what students already know,
- cognitive effort is used to make sense of ideas rather than to discover them unguided and
- knowledge is learnt as both information and through tangible examples – what we referred to as portrayal in Chapter 9.

Duration

- 10 minutes – although there will be further teacher explanations throughout the lesson.

Linking to theory

- As we saw in Chapter 4, learning science is hard for a number of reasons. This means that if we are not careful, we will overload students' limited working memory, which we learnt about in Chapter 5, when introducing new ideas. To avoid this, ideas can be introduced slowly, using concrete examples and clear explanations that use both words and graphics.

Many science lessons start with a definition and so I shall do the same as I introduce you to the powerful idea of this chapter: 'there is a certain quantity, which we call energy, that does not change in the manifold changes which nature undergoes' (Feynman on energy, taken from Millar, 2014, p. 45). Now, unless you already know what energy is, I am guessing this definition wasn't particularly helpful. Instead, I should have started with an example.

But what example can we use to introduce an abstract idea such as energy? If I was teaching Shakespeare's Song of the Witches, 'Double, double toil and trouble', I would explain what a cauldron is by making reference to a giant cooking pot. Using this method is much harder in science because there is often no tangible reference point. Energy is a scientific concept, and like all scientific concepts, is abstracted away from the physical world that we can feel, smell, see and touch. It seems like science teachers are in an impossible position, because as soon as we try to explain what a scientific concept is, we are in danger of removing the very essence that makes it a scientific concept in the first place – its abstractness.

This though doesn't mean we shouldn't try. Just that we need to tread carefully, remembering that it is through abstraction that scientific concepts gain their explanatory power (Chambers, 1991). For example, by understanding the abstract quantity of energy we know that whatever happens in the natural world – balls may roll, explosives may go bang and babies will be born, the total amount of energy is always conserved – it cannot be destroyed and you can never end up with more energy than you started with.

Introducing scientific ideas

We need then to devise alternative ways in which to portray these abstract concepts. One way or another, this involves initially making the abstract idea more concrete and tangible, allowing the idea to be experienced by our senses first before we move to the more abstract way of thinking. This chapter then will look at ways in which we can do this.

We could, for example, look to the observable, empirical phenomenon that motivated scientists to formulate their abstract ideas in the first place – this might involve seeing how objects with different masses fall at the same rate, or that crossing a smooth-seeded plant with a wrinkled-seeded plant doesn't produce semi-wrinkly seeds. Whatever example you use, you are starting with everyday observations and then moving to the scientific ideas that explain them. In the case of energy, this could involve seeing how a ball suspended from the ceiling by a wire and then held against the knee of a brave volunteer, will not hit them once released (https://spark.iop.org/massive-pendulum).

A second way is to represent the idea using words alongside models, pictures, symbols, videos and gestures. For example, we might start off by explaining how energy transfers between two objects by making comparisons to how water flows between two reservoirs located at different heights.

A third way might look to confirm the scientific idea through experimentation – this might involve predicting what happens to the mass of two substances when reacted together or seeing if the calorific content of food printed on packets accurately predicts how much energy is transferred when it is burnt in air.

In reality, of course, one explanation may use one or more of these 'three bridges' (Figure 12.1) to help students to see how the scientific ideas relate to the physical world.

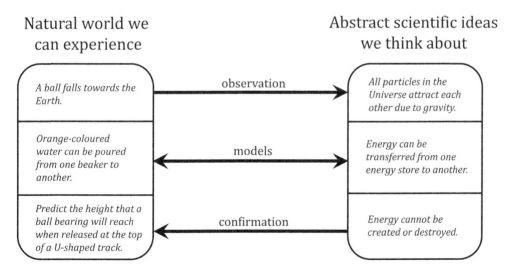

Figure 12.1 The three bridges that connect abstract ideas to the tangible natural world

▓ Teacher explanations

Good teacher explanations begin by building upon what students already know. To that end, let's set up an encounter that takes the unfamiliar scientific quantity of energy and connect it to something that students are likely to know something about. Let's burn a familiar energy resource, perhaps a Pringle, under a boiling tube of water and observe what happens with a little help from a thermometer.

Depending on how old you are, you may say that chemical energy in the Pringle is converted to thermal energy in the water. Whilst you're not wrong, this model sees energy as being transformed from one type of energy to another and so sees chemical energy in the Pringle as being different from the thermal energy in the water. But energy is energy – there are not different types of energy. Instead, a more useful way of thinking about energy is as a quasi-material substance, that is, seeing energy as an invisible substance that can flow and is the same wherever it is.

Sure, this is a simplification as energy is an abstract, mathematical quantity that needs to be calculated, but it is a useful stepping-stone with which to initially explain energy in a tangible manner that doesn't pose too many problems for later learning (Duit, 1987).

Simplify explanations at first

Sticking with our example above, of the Pringle burning beneath a boiling tube of water, we can think about ways in which energy is located or stored. There are typically eight energy stores that you will come across which I've listed below, along with events where energy may be gained or lost by that store (Fairhurst, n.d.).

- Kinetic store of energy – changes when something speeds up or slows down
- Chemical store of energy – changes when substances combine or separate
- Thermal store of energy – changes when an object's temperature goes up or down, or during a state change
- Gravitational store of energy – changes when an object has been moved in a gravitational field
- Elastic store of energy – changes when an elastic object is stretched or compressed

- Electromagnetic store of energy – changes when magnets or electric charges are pulled apart or moved closer together
- Vibration store of energy – changes when the amplitude of a mechanical wave is increased or decreased
- Nuclear store of energy – changes when nuclear particles are rearranged.

Taking a look at this list of stores, we see that before burning, energy was in the chemical store of the Pringle. As the Pringle burnt, energy was transferred from the chemical store of the Pringle to the thermal store of the water. We can represent this using a bar model (Figure 12.2), a useful representational tool for any process involving quantities, by focusing on the start and end points of our 'energy story'.

Using this diagram, we can stress the important principle that there is the same amount of energy at the end as at the beginning, simply it has transferred from one energy store to another. We have just demonstrated an important law that is the focus of the powerful idea of this chapter – that energy is always conserved.

Before Pringle burns   **After** Pringle burns

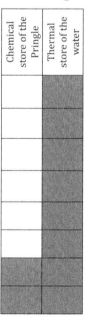

Figure 12.2 A bar model to represent how energy is conserved as it moves from one store to another

Build your explanation in layers

Those eagle-eyed amongst you will have seen that I haven't been entirely honest here.

First, we have only considered the Pringle as a store of energy, a Pringle on its own though won't do much. Instead, we need to think about the chemical store of the Pringle-oxygen system as it is only when these substances react that they release their energy (Millar, 2014). So, when we talk about energy released by burning, we need to consider the fuel-oxygen system and so Figure 12.2 will now need updating!

Second, not all energy from the Pringle-oxygen system will be transferred to the water, some will also end up in the thermal stores of the surroundings and the glass boiling tube. Eventually of course, all of the energy transferred from the Pringle will end up in the surroundings. Energy then has a natural tendency to spread out and become dissipated (anyone who has had the unfortunate experience of a pipe bursting in their home knows water behaves similarly and gets everywhere). This is why we need to switch off lights when we are not using them, not because energy is used up, but because energy spreads out, becoming much less useful. In contrast to this dissipated energy, concentrated stores of energy like coal, gas and oil are useful stores and that's why conserving these resources is so important.

Questioning to promote connections

How did energy get from the Pringle to the water?

So far we have only thought about the beginning and end of our energy story. We haven't considered how energy moves, but this must happen, otherwise how would there be a change in the way energy is stored? To answer this question, we need to think about ways in which energy is carried between stores.

There are four ways in which energy is carried (transferred) between stores and these are called pathways (Institute of Physics, n.d.). Energy can be transferred in the following ways:

- mechanically (by a force pushing or pulling something over a distance)
- heating by particle motion (because of a temperature difference)
- electrically (when a charge moves through a potential difference)
- by radiation (by electromagnetic waves)

Energy in stores is measured in joules, whereas energy carried by pathways is routinely measured in watts – the unit of power.

To understand what a watt is simply substitute units into the equation below.

$$\text{power} = \text{energy transferred}/\text{time}$$

$$P = \frac{E}{t}$$

$$W = \frac{J}{s}$$

One watt then is the transfer of 1 joule of energy per second. A quick look at the power rating of appliances in your home will now make sense. A 100 watt light bulb is transferring (not using) 100 joules of energy every second. When energy gets transferred, work is done. You can contrast this to a typical electric kettle with a power rating of 1,200 watts. Or, perhaps more interestingly, to an average Google (2009) search query that transfers around 1,020 joules, equivalent to running a 60 watt light bulb for 17 seconds.

Diagrams to focus attention

In the case of the burning Pringle, energy was transferred by heating from the chemical store of the Pringle-oxygen system to the thermal store of the water. We can represent this energy transfer using diagrams to identify only the important parts (Figure 12.3). Where possible use the teacher whiteboard and visualiser for diagrams so that you build up the explanation in layers as opposed to 'dumping it' all in one go on the screen. This allows students

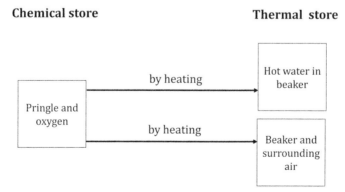

Figure 12.3 Clear diagrams alongside spoken explanations allow students to focus on key parts of an explanation, in this case how energy is transferred from one store to another when a Pringle burns in air

to see your thinking as it develops step-wise and, because you are building explanations slowly, you can be much more responsive in real time to what your students are finding tricky.

Analogy models

So far we have introduced energy by using an analogy model, where energy was compared to a quasi-material substance that could flow from one store to another. We can develop this model further by representing stores as beakers and using an orange-dyed fluid to represent energy. We can then model how energy is transferred from one store (beaker) to another by pouring liquid between them. We can even spill some of this 'energy' as we transfer the water between stores with a shaky hand as this will represent dissipation. If you collect the spillage, perhaps by carrying out the demonstration over a plastic tray, you can calculate percentage efficiency with the help of a measuring cylinder.

$$\text{Efficiency } (\%) = \frac{\text{useful output energy transfer}}{\text{total input energy transfer}} \times 100$$

Most models that you meet in science like this are analogy models; whether they are objects, pictures, equations or simulations, they all share some but not all features with the scientific concept. In terms of our energy model, the liquid can flow and be stored like energy, it's different though to energy in that energy is measured in joules not cm^3 and is not made from particles. A useful analogy model is one that has some attributes that map clearly onto the scientific idea, often referred to as the target. The best models though are not the ones that contain the most target-analog maps, but rather those that explain specific and often troublesome aspects in the best way. This means that you will want to use a number of different models to teach different parts of a concept.

The problems of models

However, whilst analogy models can be very helpful, they are a double-edged sword at times. For many students, the analogy model can soon become the reality and so we need to be careful to prevent students from taking these representations too literally. Otherwise, before we know it, atoms will become mini solar systems, electrons little balls and energy an orange-coloured liquid.

By spending time exploring how the target concept and analogy model are similar and different, you can avoid students getting too attached to any

Table 12.1 Comparing the scientific idea to the analogy model

Scientific (target) idea – a transverse water wave	Analogy model – a stadium wave
Water particles are the medium	People are the medium
Water waves are transverse with water molecules moving up and down at 90° to the direction of travel	People move up and down at 90° to the direction of travel
Energy is transferred but not matter	People move up and down but do not change seats
The wave is propagated by molecular interactions	
	The stadium wave is spread by people responding to visual cues

one representation (Coll *et al.*, 2005). The table above looks at how a stadium wave can be used to help students understand that waves transfer energy and not matter (Table 12.1). This table is completed for you, but students could spend time completing this type of activity.

Help students to also consider the purpose behind the model. This allows them to recognise how the quality of a model is not determined by how well it resembles reality, but instead how well the model helps us to reason and think about certain scientific ideas. Scale models of cells need to show how small cells are, they don't need to help us understand that the cell membrane is semi-permeable. Getting students to come up with their own model is another way in which you can limit the likelihood of them seeing the model as the real thing (Spier-Dance *et al.*, 2005).

Scientific models

Models are not just important for teaching science, though; they are also an important part of what science is, and scientists working at the cutting edge today are busy refining and generating new models to explain the world. As new scientific models become accepted, there is a growing graveyard of historical models that we still use in the classroom because they are helpful in both learning about scientific ideas and portraying the nature of science. Whilst teaching models and science models may seem quite distinct from each other in terms of their purpose – one is helping students to understand scientific concepts and one is helping to generate knowledge new to science – there is an important similarity. Both scientists and students reason through models (Hardman, 2017), allowing them to predict, explain and communicate about specific aspects of the world. Models then are not just representations of one idea, they also provide a framework with which to reason and talk about many other ideas. So, whenever I am asked a question about energy, I will often return to thinking about beakers filled with orange-coloured water.

Explicitly teaching vocabulary

The abstract quantity of energy is not only communicated through models, but also described using a vast, often unfamiliar language. Teachers of science then need to be teachers of a language too.

A particular problem with teaching scientific vocabulary is that scientific words often have different meanings from how they are used in everyday life. This is particularly problematic when teaching ideas like energy because the scientific ideas are only subtly different from their everyday meaning. It is much harder to get confused between animal cells and prison cells than between energy as something that gets transferred and energy as something that gets used up, although both misunderstandings do happen.

To think further about language in science, let's take a closer look at a definition of specific heat capacity that many students in England will need to learn.

The specific heat capacity of a substance is the amount of energy required to raise the temperature of one kilogram of the substance by one degree Celsius (AQA, 2019, p. 20).

Despite this slightly drab definition, specific heat capacity is quite a wonderful concept – it explains why cheese on your pizza burns you and why sweating water cools you down. At its most fundamental, specific heat capacity describes how much energy must be transferred to a substance to raise its temperature. Water has a relatively high specific heat capacity (4,200 J/kg °C), so it stores quite a bit of energy without a big temperature rise. This is why you can place a balloon partially filled with water above a candle without it popping and why water is such a good environment for many organisms – especially during hot summer days – as its temperature doesn't fluctuate much despite large external temperature changes.

To understand the concept of specific heat capacity, there are a number of other concepts – represented by words – that need careful introducing, and not all of these words are scientific.

A helpful way to think about introducing words in science is to start by considering the three types of words that students are likely to meet: Tier 1, Tier 2 and Tier 3 (Beck *et al.*, 2013). Tier 1 words are those that we use in everyday speech such as dog, walk and fast. Students tend to know these words without any instruction. Tier 2 words are found in exam papers or textbooks, and include words such as amble, functioning, structure and

Table 12.2 Tier one, Tier two and Tier three words needed to understand specific heat capacity

Tier one	Tier two	Tier three
The	Required	Energy
Of	Amount	Temperature
To	Raise	Kilogram
One		Substance
		Degree Celsius

evaluate. These words are likely to be encountered by students in written form but are much less likely to be used during conversations. Tier 3 words are specific to a discipline, in our case science, and are what we tend to think about when introducing new ideas. Returning to the definition of specific heat capacity, we can try to classify each word according to its tier (Table 12.2).

The specific heat capacity of a substance is the amount of energy required to raise the temperature of one kilogram of the substance by one degree Celsius.

The purpose of thinking about words in this way is it encourages us to think about both the scientific vocabulary but also about the Tier 2 words that students need to understand to make sense of the scientific idea. Tier 1 words are generally picked up in general life and don't need teaching explicitly. Tier 2 words can often be taught by referring to other words with similar, but perhaps more precise meanings. The Tier 2 word 'raise', for example, can be explained by referring to the word 'increase'. For Tier 3 words, we need to draw on a range of strategies. Below we shall look at two ways in which new words, both Tier 2 and Tier 3, can be introduced.

Word morphology

It's often helpful to spend time thinking about the morphology of words. This means looking at words in more detail and seeing if they can be broken down into smaller parts that carry meaning on their own, such as roots, prefixes and suffixes. For example, when teaching how energy is carried by electromagnetic waves, you will want to introduce students to the different regions of the electromagnetic spectrum (Radio waves, Microwaves, Infrared, Visible light, Ultraviolet, X-rays and Gamma rays). As you introduce these words you can break some of them down further to help students understand their meaning. For example, knowing the prefix 'ultra-' (from Latin) means beyond gives some meaning to the term ultraviolet – the region of the elec-tromagnetic spectrum just beyond the violet portion of the visible spectrum

Ultra-	Infra-
meaning beyond	meaning below
ultramodern	**infra**structure
ultracool	**infra**sonic
ultrasound	**infra**red
ultraviolet	

Figure 12.4 Teaching word morphology helps students to understand many new words

(Figure 12.4). Similarly, knowing that the prefix 'infra-' means below is helpful to understand that the infrared region of the spectrum corresponds to the region just below the red portion of the visible spectrum that we can see. Indeed, it was by placing a thermometer just below the red portion of a spectrum, in a region apparently devoid of sunlight, and recording a higher temperature than in any of the coloured sections, that Sir William Herschel first discovered infrared radiation in 1800.

The value of teaching word morphology is that not only does it help students to remember specific words but it allows them to make sense of a whole host of other words that share the same prefix or suffix. For example, the term infrasonic now makes more sense – a term that refers to sounds *below* the frequency that we can normally hear.

Frayer model

A second method for teaching vocabulary uses something called the Frayer model (Figure 12.5). This relies on using the word in a variety of different

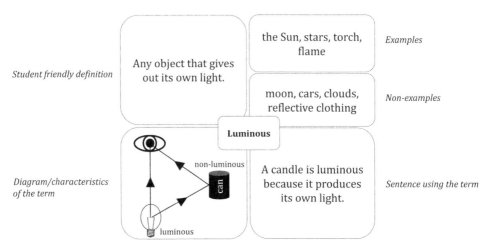

Figure 12.5 The Frayer model introduces crucial new words using examples and non-examples

ways using examples and non-examples. In this case, I have introduced the word luminous, which students may learn when studying light.

Challenges

Rather than use a non-example in this chapter, it perhaps is most helpful if we just consider what might not work and why when introducing new ideas. One of the common problems during the introduce phase is that the scientific ideas can sit, almost like a foggy cloud, above the class. The teacher is explaining ideas, often really clearly, but the students are not thinking about them because they are not clear on what these ideas mean or relate to. This can be helped by a hook that triggers situational interest, but it's a reminder that we must build explanations starting with what students already know.

Another mistake is to create tasks where students spend prolonged amounts of time gathering information. This often involves activities where students move around the classroom extracting information from summaries. The problem here is that it is very easy for students to copy, and because they are new to learning these ideas, they are not clear on what the important and interesting ideas are. It's like going weeding with an eager ten-year-old – before you know it, they've pulled up all the plants you wanted to keep. So, as a general rule, try to make the challenge in the lesson around thinking about the concepts rather than finding the information in the first place.

Taking it further

- Change where the introduce phase sits in the lesson. For example, students begin the lesson by setting up a practical to observe some phenomenon, before they are introduced to the explanation afterwards. Research shows that having children solve unfamiliar problems before instruction can improve conceptual learning (Schwartz *et al.*, 2011).
- Try having no PowerPoint slides and only use the teacher whiteboard or visualiser. You may find this helps your explanations by making them more responsive as you can respond in the moment to what students are finding easy and hard.

Summary

Scientific ideas are abstract by their very nature and so need to be introduced carefully using a range of different portrayals, such as practical demonstrations, diagrams, analogy models and explicit vocabulary instruction. All of the ideas in this chapter rely on explicit teacher instruction, supported by deep subject knowledge that guides students to see through the complexities of the new information, using careful questioning along the way. However, for all the positives, there is one major drawback to explicit teacher instruction. Explicit teacher instruction requires all students to move at the same rate. Whilst this can be maintained for ten or so minutes, the longer this goes on, the larger the gap becomes between those who are following and those who are lost. This means we need to create opportunities for students to move at their own rate. This involves creating tasks where students practice using their knowledge for sustained periods of time and so this is where we head to next.

Bibliography

AQA. 2019. GCSE physics specification v1.1. Available at: https://www.aqa.org.uk/subjects/science/gcse/physics-8463. [Accessed: 15 December 2019.]

Beck, I. L., McKeown, M. G. and Kucan, L. 2013. *Bringing words to life: Robust vocabulary instruction*. New York, NY: Guilford.

Chambers, J. H. 1991. The difference between the abstract concepts of science and the general concepts of empirical educational research. *The Journal of Educational Thought/Revue de la Pensée Educative*, 25, pp. 41–49.

Coll, R. K., France, B. and Taylor, I. 2005. The role of models/and analogies in science education: Implications from research. *International Journal of Science Education*, 27(2), pp. 183–198.

Duit, R. 1987. Should energy be illustrated as something quasi-material? *International Journal of Science Education*, 9(2), pp. 139–145.

Fairhurst, P. n.d. *Approaches to teaching energy*. Best Evidence Science Teaching Resources. Available at: https://www.stem.org.uk/sites/default/files/pages/downloads/BEST_Article_Teaching%20energy.pdf. [Accessed: 15 December 2019.]

Google. 2009. *Powering a Google search*. Available at: https://googleblog.blogspot.com/2009/01/powering-google-search.html. [Accessed 15 December 2019.]

Hardman, M. A. 2017. Models, matter and truth in doing and learning science. *School Science Review*, 98(365), 91–98.

Institute of Physics. n.d. *Helpful language for energy talk*. Available at: https:// spark.iop.org/helpful-language-energy-talk. [Accessed: 19 May 2020.]

Millar, R. 2014. Teaching about energy: From everyday to scientific understandings. *School Science Review*, 96(354), pp. 45–50.

Schwartz, D. L., Chase, C. C., Oppezzo, M. A. and Chin, D. B. 2011. Practicing versus inventing with contrasting cases: The effects of telling first on learning and transfer. *Journal of Educational Psychology*, 103, pp. 759–775.

Spier-Dance, L., Mayer Smith, J., Dance, N. and Khan, S. 2005. The role of student generated analogies in promoting conceptual understanding for undergraduate chemistry students. *Research in Science & Technological Education*, 23(2), pp. 163–178.

CHAPTER 13

Practice to build understanding
Worked examples and deliberate practice

Powerful idea focus

- *Quantities in chemistry are expressed at the macroscopic and submicroscopic scales using grams, volumes and moles.*

Pedagogy focus

- Practicing conceptual and procedural knowledge.

Aims

The practice phase provides an opportunity for students to:

- practice using the knowledge from the introduce phase, for example through answering questions, labelling diagrams or using a specific piece of apparatus,
- practice specific aspects of solving a problem, which progress in difficulty and range and
- get one-to-one support from the teacher if they are finding it hard.

Duration

- 10–15 minutes.

▮ Linking to theory

● This phase looks at the role of deliberate practice to develop expertise by breaking down complex problems into smaller steps that can be practiced separately. This links to what we learnt in Chapter 5 about the need to reduce extraneous cognitive load. We also look at how to structure practice using worked examples and how practice might look different when learning conceptual and procedural knowledge.

If you were to ask yourself how scientists think differently to students you would probably come up with ideas such as:

● scientists know more,

● scientists generate knowledge new to science and

● scientists know how to use complicated pieces of apparatus.

Perhaps though, the key difference is scientists are just better at jumping.

When students are presented with a chemical formula such as $CuSO_4$ (aq), they might just see a collection of symbols that correspond to the elements copper, sulfur and oxygen. A scientist will see these symbols too, but they may also visualise a beautiful blue solution made up of copper and sulfate ions jostling about separately as they form new interactions with water molecules. In other words, as we saw in Chapter 4, scientists are particularly good at thinking about scientific concepts at multiple levels, easily moving between them as they come to describe and explain scientific phenomena; however, beginners find this much harder (Figure 13.1).

Experts are able to think about these different levels because they understand that a word or symbol can be thought about in different ways and so

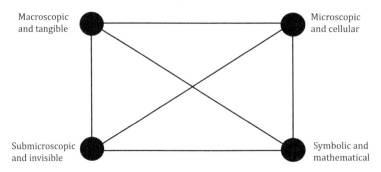

Figure 13.1 Experts are good at jumping between these levels when they think about scientific ideas; students will need some help

can connect the descriptive, macroscopic levels to the explanatory submicro-scopic, mathematical and cellular levels (Taber, 2013). This 'jumping' is very demanding when you are first learning science because it's not always clear how ideas at one level relate to ideas in another. How, for example, does universal indicator going green relate to the movement of ions inside the conical flask? Neither is it always clear as to which level you should use to explain your answer. For example, in a physics lesson, explaining why Granny slipped on ice by referring to a loss of muscle strength is unlikely to score many marks, despite this being a valid explanation. Instead, physicists will want you to talk about friction and other forces. This can leave students learning science in a quandary – how do they know what explanation is appropriate, when?

The psychologist K. Anders Ericsson sees expertise like this as developing, not through experience, that is, exposing students to all levels at once and hoping that they will pick it up by osmosis (I'm not referring to the scientific meaning here), but rather through something called deliberate practice (Ericsson, 2008). Deliberate practice is a special form of practice where students:

– complete a task with a clearly defined goal,
– are motivated to improve,
– are given feedback and
– have opportunities for repetition and gradual refinement of performance.

In science, this involves taking a complex problem and breaking it down into discrete steps that students then practice and tends to focus on one of the levels described above in the short-term. Over time we can then create tasks for students to practice making connections between one level and another. The focus of this chapter is to explore how we use the idea of repetition and refinement to help students build an understanding of the powerful idea – that *quantities in chemistry are expressed at the macroscopic and submicroscopic scales using grams, volumes and moles.*

The mole

At the heart of this powerful idea is the fact that we can take any substance and work out how many atoms or ions it is made from, simply by measuring its mass if it is a solid or liquid, or measuring its volume if it is a gas* or solution. So thanks to the concept of a mole, we know that in 23.0 grams of sodium, or indeed the mass of one mole of any substance found using

---

* Note that the molar volume of a gas is affected by temperature and pressure.

the periodic table, we have approximately 600 billion trillion individual particles. This is an enormous number, and, to put it into context, it is more red blood cells that exist in every human being right now (Eveleth, 2013).

The mole has been a base unit of the International System of Units (abbreviated to SI, from the French Système international d'unités) since 1971 and can be quite a tricky idea for students to learn. At its most fundamental, a mole is an amount of stuff. In the 1970s, this amount of stuff was defined as the amount of substance that contained as many elementary entities as there were atoms in 0.012 kilograms of the most common form of carbon, known as carbon-12. This definition had a problem though, because a mole was understood in relation to a kilogram which was understood in relation to a block of metal called 'Le Grand K', and, like most physical objects, Le Grand K was changing. This meant, based on the definition given above, so too was the mole. To get around this problem, one mole was recently redefined as containing exactly $6.02214076 \times 10^{23}$ elementary entities (Meija, 2018) and so no longer relies on a physical object.

A mole then is just a 'label' like a dozen or a pair. This label refers to a specific number of particles that we can't see, but it can also describe a specific mass or volume of a substance that we can see. For example, one mole of water has a mass of 18.0 grams comprised of 6.02×10^{23} water molecules, made from 1.81×10^{24} atoms. This means that when we talk about one mole of water, or indeed one mole of any substance, we need to be clear about exactly what we are referring to – that is, are we talking about a mass or a number of atoms, molecules or ions (Figure 13.2).

We can use this number, 6.02×10^{23} – referred to as Avogadro's number – to count in chemistry. Imagine, for example, that we wanted to react iron with sulfur to make iron sulfide with no atoms of iron or sulfur left over.

$$Fe + S \rightarrow FeS$$

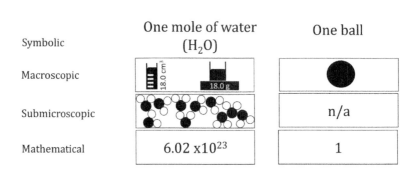

Figure 13.2 One mole of water looks very different depending on what you are referring to

Weighing equal masses of each substance is not going to work because an atom of iron is much heavier than an atom of sulfur – nearly twice as heavy if you consult the periodic table. Instead, what we need to do is react one mole of iron with one mole of sulfur because, as we've seen above, one mole of any substance will contain the same number of particles – 6.02×10^{23}. In this case, we simply weigh out 56.0 grams of iron and react it with 32.0 grams of sulfur – that is, we simply weigh out, in grams, their relative atomic masses that are found on the periodic table.

Breaking down complex problems

One of the most powerful aspects of the mole is that it allows us to make very accurate predictions.

Imagine, for example, we are on a desert island, washed up with a barrel containing 50.6 grams of 100% ethanol. We know not to drink this ethanol because its purity will kill us but wonder whether we can react this ethanol with oxygen to produce enough water to drink. To help us decide, we will need to start with a chemical equation.

$$C_2H_5OH + O_2 \rightarrow CO_2 + H_2O$$

This equation is not balanced, though – there are more atoms of carbon in the reactants (left-hand side) than in the products. Hydrogen atoms are not balanced either. This situation is clearly impossible because as we saw in Chapter 6, we cannot destroy or create atoms (at least in a test tube we can't) – we can only rearrange them.

If we were going to teach students how to balance this equation, we need to first figure out *exactly* what students need to know and be able to do to solve this type of problem. The best way to work this out is to complete the task yourself, otherwise we are in danger of assuming it is easier than it really is – a dangerous condition known as expertise-induced blindness that all teachers suffer from!

As you balance the equation below, write a list of everything you needed to know and do to solve it.

$$C_2H_5OH + O_2 \rightarrow CO_2 + H_2O$$

What students need to know and do to balance an equation

1. Symbols represent elements and these are found on the periodic table.
2. Substances are made from atoms chemically bonded together.

3. Atoms join together in fixed proportions to form substances.
4. Substances can be represented by formulae showing relative numbers of atoms present.
5. Formulae can refer to amounts of substance or individual atoms.
6. A number within a formula means multiples of the atoms to the left of the number.
7. On the left of the arrow lie the reactants on the right lie the products.
8. The arrow represents a chemical reaction where a new substance is made.
9. When this reaction has taken place there are no reactants left.
10. Conservation of mass:

    i. Atoms can be rearranged in a reaction.
    ii. Atoms cannot be destroyed in a reaction.

11. To change the number of atoms you add a coefficient before the substance.
12. We assume a coefficient of 1 before each formula if nothing is written.

So, nested within what appears to be quite a simple operation of balancing an equation, lies an enormous amount of knowledge that students are going to need to practice using and thinking about before they attempt to solve the problem as a whole.

Procedural and conceptual knowledge

Knowledge in the list above can be placed into one of two camps – procedural knowledge and conceptual knowledge. Thinking about these two types of knowledge can be helpful because it affects how we will design the practice activities.

Conceptual knowledge, as the name suggests, is knowledge of concepts. Knowing for example that substances are made from atoms chemically bonded together is an example of conceptual knowledge. Procedural knowledge is different and involves knowing how to carry out specific procedures by following a series of steps, for example, knowing how to balance an equation by adding coefficients before the formula.

Whilst there is an important interplay between these two types of knowledge, it is perfectly possible to balance a chemical equation without having any idea about the underlying chemical concepts of substance, formula, reactions or conservation of mass, and instead simply follow a series of rules (Nyachwaya *et al.*, 2014).

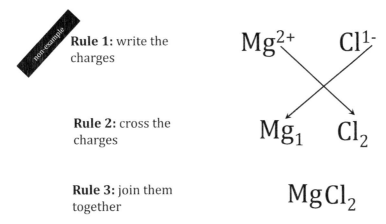

Rule 1: write the charges

Rule 2: cross the charges

Rule 3: join them together

Mg^{2+} Cl^{1-}

Mg_1 Cl_2

$MgCl_2$

Figure 13.3 Be careful asking students to follow rules that experts don't use

The problem here is not rules *per se*, just that they need to be the right kind of rules. To illustrate what I mean let's take a brief detour to look at how to construct the formula for an ionic compound such as magnesium chloride. It is perfectly possible to teach students how to do this using a 'crossing method' (Figure 13.3). This procedure gets students to the right answer by following rules; however, it is not how an expert would solve this problem as it does not explicitly consider balancing charges and so is storing up problems for later. Students who have learnt how to write ionic formula using these rules would be completely stuck if asked to write the formula for sodium hydrogen carbonate ($NaHCO_3$)! Rules in lessons, then, should try, as much as possible, to articulate how a subject expert would solve the problem and not simply be geared towards getting the correct answer. This is why students should learn how to rearrange equations and not use formula triangles.

Worked examples

Returning to balancing equations, let's consider how worked examples help students to solve complex tasks. As the name suggests, this involves first showing students how to solve a problem before getting them to solve it on their own. After the worked example, students get time to practice a number of related questions and get feedback. The worked example makes our thinking about how to carry out procedures explicit and available to the students, who can then use this method to solve other problems. Worked examples reduce extraneous cognitive load compared to solving problems without any guidance – and this is called the worked example effect (Sweller *et al.*, 2011).

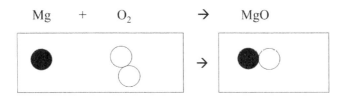

Figure 13.4 Particle diagram for the unbalanced equation

Let's use a worked example to illustrate how to balance a chemical equation – the reaction of magnesium with oxygen – but importantly you can take these principles and use them to teach how to solve many other procedural problems in science. Whilst the worked example is set out below in separate steps, students find it helpful if they are built up in layers using the whiteboard or visualiser. This means you would only need to draw one table and then simply add to it. As you move through the worked example, speak aloud your thoughts as it's your thinking that students really need to hear.

Step 1: write out the equation and draw a diagram to show the particles present (Figure 13.4). This makes the point that the equation is not balanced. Students will need some separate practice drawing pictures like these before they balance equations.

Step 2: use the equation to create a table and identify the *elements* present (Table 13.1).

Table 13.1 Use the equation to create a table and identify the *elements* present

	Mg	O_2	→	MgO
Mg				
O				

Step 3: record the number of each type of atom/ion and check if it is balanced (Table 13.2).

Table 13.2 Record the number of each type of atom/ion and check if it is balanced

	Mg	O_2	→	MgO	Balanced?
Mg	1			1	Y
O	2			1	N

Step 4: add a coefficient **before** a substance to balance the atoms/ions. In this example, we will balance oxygen. If you have many elements to balance,

a good tip is to leave oxygen until the end as it tends to be a loner as a reactant. Update the table (Table 13.3).

Table 13.3 Add a **coefficient** before a substance to balance the atoms/ions; update the table

	Mg	O_2	→	2MgO	Balanced?
Mg		1		2+	~~Y~~ N
O		2		2+	~~N~~ Y

Step 5: you will now need to repeat step 4 to balance other atoms. In this example, we will now balance magnesium. Update the table (Table 13.4).

Table 13.4 Repeat step 4 to balance the atoms; update the table

	2Mg	O_2	→	2MgO	Balanced?
Mg		2+		2+	~~Y~~ ~~N~~ Y
O		2		2+	~~N~~ Y

Step 6: when all atoms are balanced, write out the balanced equation and update your diagram in step 1.

$$2Mg + O_2 \rightarrow ......$$

A danger of worked examples is students simply copy, and don't pay attention to thinking about what is actually going on. This would make it difficult to reproduce the same performance at a later stage – a bit like blindly following satnav, arriving at your destination but having no clue of how you got there or how to get home – we've all been there. To get around this problem, you can present the worked example but with some information missing. In the example above, step six was left incomplete. If students understand the reasoning behind the worked example, they should be able to see what the final answer is.

Students will now need to balance a number of similar equations that progress in difficulty but not so difficult that they can't be solved using the worked example:

- $Ca + O_2 \rightarrow CaO$
- $C + H_2 \rightarrow CH_4$
- $Na + Cl_2 \rightarrow NaCl$

- $H_2 + O_2 \rightarrow H_2O$
- $H_2 + Cl_2 \rightarrow HCl$
- $KBr + F_2 \rightarrow KF + Br_2$
- $NaI + Br_2 \rightarrow NaBr + I_2$
- $Al + Cl_2 \rightarrow AlCl_3$
- $CH_4 + O_2 \rightarrow CO_2 + H_2O$
- $H_2SO_4 + NaOH \rightarrow Na_2SO_4 + H_2O$
- $C_2H_5OH + O_2 \rightarrow CO_2 + H_2O$

Using equations

Now that students can balance equations, it's time for us to return to our desert island to see their predictive power!

If we have 50.6 grams of ethanol in our barrel, then according to the equation below, we have 1.1 moles of ethanol.

$$\text{number of moles (mol)} = \frac{\text{mass (g)}}{\text{molar mass (g/mol)}}$$

$$\text{number of moles (mol)} = \frac{50.6 \text{ (g)}}{46.0 \text{ (g/mol)}}$$

$$\text{number of moles (mol)} = 1.1$$

Using the stoichiometric ratio from our balanced equation ($C_2H_5OH + 3O_2 \rightarrow 2CO_2 + 3H_2O$), we will make three times as many moles of water as we have ethanol. In that case, we would make 3.3 moles of water, and because we can calculate the mass of one mole of water using the periodic table (18.0 g), we can rearrange the equation used above to find out the mass of water we would make – a rather impressive 59.4 grams which is certainly enough to drink! And to help students see how we might go about doing this reaction, you might want to demonstrate the 'whoosh bottle'* (https://edu.rsc.org/resources/the-whoosh-bottle-demonstration/708.article).

Checking procedures make sense

So far, we have focused on procedural knowledge – the ability to do something such as how to balance equations or calculate how much of something we can make. But as we saw earlier, it is perfectly possible to balance chemical

---

* Please don't drink the products from this demonstration.

equations without understanding *why* equations need to be balanced in the first place, or what these letters and symbols actually relate to. This means that practicing just procedural knowledge is insufficient. We also need to help students to develop conceptual knowledge so that the rationale for the procedures they use is understood. To explore what this could look like, let's use a few examples to illustrate how we might help students think about some of the reasons behind the procedures they are carrying out when balancing equations. These examples could be applied to any topic area and often involve explaining why right is right or why wrong is wrong.

Make valid comparisons between seemingly unrelated procedures

If you understand the purpose behind a procedure, it is likely that you can make comparisons between seemingly unrelated processes that share a deeper purpose. In the question below, students are asked to apply their understanding of balanced equations to a different but related context. Hopefully they will see that equation 1 below is the odd one out as it is the only balanced equation.

Identify which equation (1-3) is the odd one out. Explain your choice.

1. frame + wheel + wheel → scooter
2. $Ca + O_2 → CaO$
3. board + wheel → skateboard

I think this because …

Spot the error in the answer and explain why

So much of science is spent trying to get the right answer. Important though a right answer is, students should also be able to spot and explain mistakes. The example below requires students to identify, explain and correct an error to an incorrectly balanced equation.

$$N_2(g) + H_2(g) → NH_3(g)$$

This equation is wrong because …
    The correct equation is …

Spot the error in the working and explain why

If you understand a procedure you can identify and explain errors in the working too. In Table 13.5, the student didn't use the balanced equation

Table 13.5 Using a table to calculate reacting masses; can you spot the error?

	C_2H_5OH	$3O_2$	$\rightarrow$	$2CO_2$	$3H_2O$
Stoichiometric ratio	1	3		2	3
Mass of one mole (g)	46				18
Actual mass you have (g)	50.6				19.8
Amount (mol)	1.1				1.1

when calculating the theoretical moles of water that could be produced by the reaction.

> What has the student done wrong?
>> Explain what the student should have done …
>>> The correct answer is …

Practicing conceptual knowledge

Unlike practicing procedural knowledge, where tasks follow the same format, practicing conceptual knowledge seeks to avoid students relying too much on task-specific characteristics (Rittle-Johnson and Schneider, 2015). Instead, the focus should be on students organising and connecting knowledge so they can appreciate the deep underlying relationships between seemingly unrelated phenomena – for example, seeing that what connects 40.1 grams of calcium, 12.0 grams of carbon and 52.0 grams of chromium are molar masses and not the fact they are all elements beginning with the letter c. This involves students completing a variety of divergent tasks where they gradually come to see the similarities and differences between the concepts they are using.

Below, we will look at three ways in which conceptual knowledge can be developed through:

- use of Venn diagrams and categorisation,
- translating ideas between different levels and
- refutation tasks.

As with all examples used in this book, they can be adapted to teach a range of topics.

Making comparisons

An important part of conceptual understanding is to understand how two concepts are similar and how they differ. We've talked about this before in

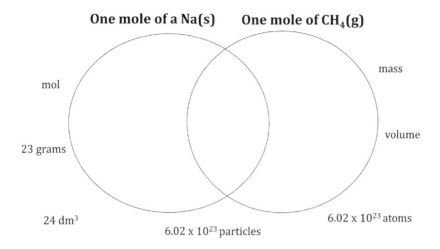

One mole of a Na(s)   One mole of CH$_4$(g)

mass

mol

volume

23 grams

24 dm^3 6.02 x 10^{23} atoms

6.02 x 10^{23} particles

Figure 13.5 Venn diagrams can help students to see the similarities and differences between concepts

terms of cats and dogs – to understand what a cat is, you need to know why a dog isn't one. Venn diagrams are an excellent way for students to practice using concepts to see how they are related. If this is done in pairs, then students can discuss and refine their ideas together. You can provide a focus by providing some concepts to be sorted (Figure 13.5). In this case, we are looking at the similarities and differences in how moles relate to gases and solids. After an exercise like this, students can be given the opportunity to write a paragraph to compare and contrast the ideas that formed the basis for the Venn diagram. For example, in this case, students could use their Venn diagram to explain how a mole of methane is similar and different to a mole of sodium; like a good lesson, the task itself builds over time.

Refutation tasks

Misconceptions are generally concerned with conceptual knowledge. Refutation tasks are a specific way to address misconceptions that we learnt about in Chapter 4. The task involves a text passage that contains a commonly held misconception and an explicit rejection of that misconception by explaining the scientifically acceptable idea (Tippett, 2010). Refutation texts appear to be most successful when misconceptions are single ideas rather than complex concepts (Beker *et al.*, 2019).

> **Refutation:** some people think that one mole of any substance will have the same mass. For example, they may think that a mole of Cl$_2$ and a mole of O$_2$ will always have the same mass.

> **Explanation:** however, this is wrong because if you look at the periodic table atoms do not all have the same mass. Chlorine atoms are heavier than oxygen atoms. One mole of Cl_2 has a mass of 71 grams and one mole of O_2 has a mass of 32 grams. However, it is true to say that one mole of any gas, at the same temperature and pressure, will occupy exactly the same volume.

You can then pose some follow-up questions after the refutation text to consolidate understanding.

Translating ideas between levels of thinking

One of my favourite ways to practice using conceptual knowledge is to get students to translate the same idea between different levels of thinking. In the example below, students take information from a question and then translate this into both macroscopic and submicroscopic levels (Figure 13.6). The example I have shown is for a titration, but you could use this principle for lots of different examples where you want students to zoom in or out at specific points in a diagram so they build a depth of understanding – that is, to see concepts such as titrations as not just about burettes but also about what is happening inside the glassware to the particles.

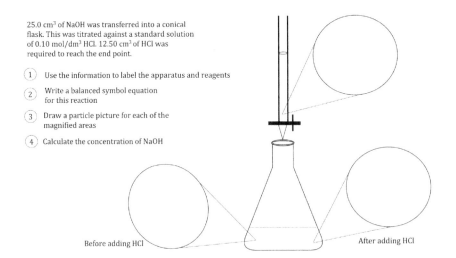

25.0 cm³ of NaOH was transferred into a conical flask. This was titrated against a standard solution of 0.10 mol/dm³ HCl. 12.50 cm³ of HCl was required to reach the end point.

1. Use the information to label the apparatus and reagents

2. Write a balanced symbol equation for this reaction

3. Draw a particle picture for each of the magnified areas

4. Calculate the concentration of NaOH

Before adding HCl

After adding HCl

Figure 13.6 Linking the macroscopic to the submicroscopic by asking students to draw particle pictures

1 mole of CO_2 has a mass of 44 g

Using what you have learnt, calculate the mass of 1 mole of the following substances

1. $MgCO_3$
2. $Mg(NO_3)_2$
3. Aluminium oxide

Figure 13.7 A non-example for the practice phase

Non-example

We've looked at lots of examples in this chapter of ways for students to practice using knowledge in science, whether it's through using worked examples, Venn diagrams or refutation texts. Take some time to look at the non-example (Figure 13.7) to identify why it might not be a good way of practicing how to calculate relative formula mass.

The non-example above may not work well as a practice task because:

- the example at the top does not show how the answer of 44 was arrived at,
- the example does not help with answering questions two or three,
- there is only one opportunity to practice each type of problem, for example, there is only one question on calculating the molar mass for formulae with brackets and
- the task is too short, meaning some students will finish before some have begun and so there is no opportunity for differentiation.

Challenges

We've already touched on some of the challenges of practice in terms of providing the wrong rules – these are shortcuts that don't resemble how an expert scientist might solve a problem. Don't be afraid of dedicating considerable time for students to practice the ideas that have been introduced and consider how practice looks different for procedural knowledge and conceptual knowledge. If the practice phase is too short, students won't have time to make mistakes and subsequently overcome them. If the practice is too easy, students won't be thinking about what they are doing. All practice needs feedback, and so when giving lots of questions it can be helpful to also give answers at points so students can self-mark as they go and develop confidence. As long as you require full working, you will still be able to distinguish those students who have understood from those who have simply copied answers.

Taking it further

- Try live marking to provide feedback. This involves moving around the class and providing brief written feedback as students complete questions.
- As the majority of the class are completing questions, you can work with a small group either to extend or to support understanding.

Summary

In some ways practice can feel the least exciting phase of teaching science. It's a time when students get down to work, often working individually for sustained periods of time and so teachers can feel like a slightly spare part. But practice is a fundamental part of the learning process, where students get to work at their own rate to make sense of ideas, make mistakes and get feedback. It is also a time where teachers can support students one-to-one if they are finding it hard. In this chapter, we have looked at ways to structure, or scaffold, tasks considering the type of knowledge students need to practice. Practicing procedural knowledge requires repetition of similar problems to develop fluency, whereas practicing conceptual knowledge requires a greater variety of tasks and opportunities for discussion. We have also considered the differences between good rules that experts follow and bad rules that simply get students to the right answer but store up trouble for later. As students become more confident throughout the practice phase, they will become familiar with the ideas within one level of thinking, for example, the submicroscopic or macroscopic, and start connecting different levels together to resemble how an expert thinks about science. The next stage is now to do something with all of this knowledge – it's time to apply and integrate these ideas.

▣ Bibliography

Beker, K., Kim, J., Van Boekel, M., van den Broek, P. and Kendeou, P. 2019. Refutation texts enhance spontaneous transfer of knowledge. *Contemporary Educational Psychology*, 56, pp. 67–78.

Ericsson, K. A. 2008. Deliberate practice and acquisition of expert performance: A general overview. *Academic Emergency Medicine*, 15(11), pp. 988–994.

Eveleth, R. 2013. *How big is a mole?* Available at: https://blog.ed.ted.com /2013/10/23/how-big-is-a-mole-exactly/. [Accessed: 15 December 2019.]

Meija, J. 2018. *A new definition of the mole*. Available at: https://www .chemistryworld.com/opinion/a-new-definition-of-the-mole/3008663 .article. [Accessed:15 December 2019.]

Nyachwaya, J. M., Warfa, A. R. M., Roehrig, G. H. and Schneider, J. L. 2014. College chemistry students' use of memorized algorithms in chemical reactions. *Chemistry Education Research and Practice*, 15(1), pp. 81–93.

Rittle-Johnson, B. and Schneider, M. 2015. Developing conceptual and procedural knowledge of mathematics, in *Oxford handbook of numerical cognition*, Kadosh, R. C. and Dowker, A. (eds.), 1118–1134. Oxford: Oxford University Press.

Sweller, J., Ayres, P. and Kalyuga, S. 2011. *Cognitive load theory*. New York, NY: Springer.

Taber, K. S. 2013. Revisiting the chemistry triplet: Drawing upon the nature of chemical knowledge and the psychology of learning to inform chemistry education. *Chemistry Education Research and Practice*, 14(2), pp. 156–168.

Tippett, C. D. 2010. Refutation text in science education: A review of two decades of research. *International Journal of Science and Mathematics Education*, 8(6), pp. 951–970.

Apply and integrate to make and break connections

▓ Powerful idea focus

- *Organisms compete with, or depend on, other organisms for the same basic materials and energy that cycle throughout ecosystems.*

▓ Pedagogy focus

- Using knowledge and skills in more unfamiliar contexts.

▓ Aims

The apply phase provides an opportunity for students to:

- use and extend their knowledge in a range of varied examples that promotes a deeper understanding,
- use the knowledge from the introduce and build phases in a way that publicly demonstrates they understand it,
- integrate the new ideas they have learnt into their world by making it their own – for example, using their knowledge and skills to plan an investigation to answer a question that matters to them and
- provide the opportunity for students to develop disciplinary literacy through talking, reading and writing in science.

▓ Linking to theory

- This phase draws on theories of motivation discussed in Chapter 6 and considers how experts – that is, those individuals with lots of prior knowledge about a specific idea – have different instructional needs to beginners.

So far in this book we have mainly considered learning as a product of teacher instruction. In other words, learning happens when teachers carefully scaffold explanations and create structured opportunities for practice. However, it would be a mistake to think that students only learn what they've been told. This, as I hope to show you in this chapter, is absolutely not the case.

As we saw in Chapters 4 and 5, learning, at its most fundamental, is about making and breaking connections. In effect, it's about realisations, moments where seemingly unrelated ideas are suddenly brought together to make sense of something else. So, whilst students may know the equations for photosynthesis and respiration, they may have failed to see the connection.

Respiration: glucose (food) + oxygen → water + carbon dioxide
Photosynthesis: water + carbon dioxide → oxygen + glucose (food)

The connection, of course, being that photosynthesis, the process by which plants make food, is respiration in reverse. Neither may students have realised that respiration is an exothermic process and photosynthesis an endothermic one. If they have made this connection, they will realise that plants perform photosynthesis *and* respiration, animals, such as us, just respire, meaning we have to go to the supermarket to get our food.

Whilst it is tempting to believe that all realisations like this can be achieved through carefully guided instruction, I don't believe this is the case. Or at least it may be the case sometimes, but not in a classroom packed full of 30-something individuals, all of which have different connections to make.

Expertise-reversal effect

As students acquire more knowledge about a scientific idea, the differences in what they know increases. This is simply a consequence of probability, because as we know more the number of possible connections we can make increases too. In other words, when prior knowledge is low, it is easier to teach students using explicit instruction methods because their needs are more similar. However, as students acquire more knowledge about an idea, their understanding begins to diverge, often in subtly different ways that are not always obvious for us to see. This means that explicit instruction, where all students are taught the same thing, starts to become ineffective for certain individuals. This is called the expertise-reversal effect and explains why many instructional tools, such as worked examples, become problematic for students as they develop their knowledge of an idea (Kalyuga *et al.*, 1998). Here, any irrelevant information, such as when you include diagrams

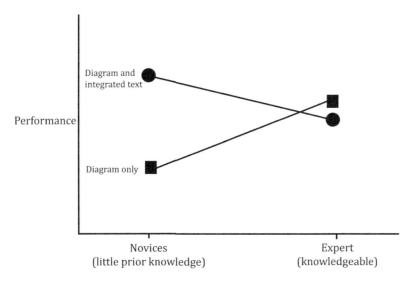

Figure 14.1 What works for novices doesn't always work for experts

Source: Informed by Kalyuga *et al.*, 1998.

and text together, simply increases extraneous load of experts that we learnt about in Chapter 5 (Figure 14.1). It's like being told how to do something when you already know how to do it – a complete distraction!

So, whilst we may initially scaffold learning carefully using explicit teacher instruction, worked examples and deliberate practice, once students have gained sufficient prior knowledge, we want to create opportunities for students to engage in more open tasks.

During activities like these, students have the opportunity to bring together what they have learnt from other related areas from both the world of science and their everyday lives. Not only does this integration play an important role in re-organising knowledge, it also plays a role in motivation by allowing students to see how their performance has improved – there is nothing more motivating than seeing what you have learnt.

With this in mind, let's explore how we can create opportunities for students to apply and refine what they know about the powerful idea of this chapter, that is, *that organisms compete with, or depend on, other organisms for the same basic materials and energy that cycle throughout ecosystems.*

We are all made from the same stuff

When you stop and think about it, it is actually pretty incredible that you, me, a mushroom, a bacterium and a blade of grass are all made from the same basic stuff. Living organisms then, irrespective of their size or habitat, are mostly made from proteins, carbohydrates, fats, water, minerals and vitamins* that themselves are made up from the elements oxygen, carbon, hydrogen, nitrogen, calcium, phosphorus and sometimes a dash of sulfur. The fact that all organisms are made from the same building blocks has some important implications in that we depend on each other for our survival. When we eat a plant, we get nitrogen, amongst other things, that we use to make proteins, but plants get nitrogen from us if they have the misfortune to be peed on or worse. Admittedly, nitrogen uptake by plants is slightly more complex than this and needs the help of many bacteria living in the soil, but you get the idea – organisms are dependent on each other. There is nothing inevitable about this – we could imagine a different world, one where humans are made from different building blocks to plants and bacteria. Perhaps a world where life evolved independently more than once and stumbled on different ways of doing things.

There is, though, a downside of being made from the same stuff as your neighbour and the species that surround you. Because all organisms are made from the same basic materials, it means we are forced to compete with one another when these materials are in short supply. Bacteria attack our cells and scavenge for the very same resources that we need. Intestinal parasites eat our food and fungi get between our toes. As we will see in Chapter 15, this means that those individuals with longer roots, sharper teeth or longer necks may be at an advantage causing the weaker individuals to lose out.

Photosynthesis

The pencil in your pocket, the cotton jumper you might be wearing and the food you had for lunch all came, either directly or indirectly, from plants. Plants are the ultimate source of all food on Earth, thanks to their ability to carry out photosynthesis where energy is transferred from a nuclear store (the Sun) to a chemical store (glucose plus oxygen). It is this glucose that we and other organism, including plants, feed on.

This overly simplified equation for photosynthesis (Figure 14.2) hides a number of complexities that many students fail to recognise (Hershey, 2004):

- Carbon dioxide comes from the air not the soil
- Glucose is not the major photosynthetic product, it's sucrose and starch
- Plants carry out photosynthesis *and* respiration

---

* We should add nucleic acids to our list too. Our bodies apparently contain about 50 grams of human DNA.

$$6CO_2 + 6H_2O \xrightarrow{\text{Light}} 6O_2 + C_6H_{12}O_6$$

Figure 14.2 Photosynthesis – the process through which plants make their (and our) food

- Much more water is needed to produce one mole of glucose than this equation shows as stomata (pores) in the leaf must be open to let carbon dioxide in, with the unfortunate loss of water out
- The one arrow makes it look like the process happens in one step, when in fact there are many reactions involved

This oversimplification is desirable when students are first learning about photosynthesis, but when the fundamental prior knowledge is in place, it's time to get students to think more deeply about some of these aspects by asking them to consider how the equation for photosynthesis relates to other scientific ideas – it's time for them to connect up that web-of-knowledge that we learnt about in Chapter 5.

In the rest of this chapter, we will look at different ways in which we can encourage students to apply and integrate their ideas using a range of open-ended tasks. The purpose of these activities is to create the opportunity for students to think more deeply about the ideas they have learnt about, refine their ideas and to reveal their thinking which, as we will see in Chapter 15, provides an important source of material for feedback.

Extended writing: using fantasy

In Chapter 6, we looked at how emotions play an important role in learning science. We're now going to look at a specific example that uses fantasy to really probe what students understand about two of life's most important processes, photosynthesis and respiration, using the story of the Green Child (Figure 14.3).

This fictitious, or fantastical story, provides a great opportunity for students to apply what they know about photosynthesis and biology to a new, yet related context and to integrate these ideas together (Green, n.d.). The questions get students thinking hard about the relationship between glucose, a product of photosynthesis, and food. Food is not simply something that gets eaten but has a more technical meaning – food is a substrate for respiration and so is analogous to a fuel. The Green Child will still need to eat because she will need to obtain minerals and vitamins, however, she won't need to eat food because she can make her own food through photosynthesis. Like all organisms, the Green Child still respires because she needs to transfer

Once upon a time, a human zygote became infected with a photosynthetic bacterium. When the baby was born she could photosynthesise. This 'Green Child' was very special because...

The Green Child

> **Answer the questions below. Then plan out how you will use these answers to continue the story.**
> ❑ What process could this child do that other human children could not?
> ❑ Why did the Green Child only feel hungry at night time?
> ❑ Did the Green Child have to eat?
> ❑ Include a labelled diagram of the Green Child's cells in your story.
> ❑ Do you think every cell in the Green Child could photosynthesise? Explain your answer.
> ❑ Would the Green Child still need to use their lungs? Explain your answer.

Figure 14.3 The use of fantasy to encourage students to apply their knowledge

Source: Drawing by Arabella Green.

energy from her food in order for muscle contraction and other endothermic processes to take place. Her cells will therefore contain mitochondria.

Once students have had the opportunity to discuss and answer these questions they can continue the story using extended writing. It helps if they are clear who the audience is – in this case their teacher – before they write. Through the process of writing, students not only display their thinking but the act of writing helps them to organise their ideas and provides the opportunity to develop their disciplinary literacy – see https://www .serpinstitute.org/reading-science for further ideas.

The role of fantasy in the story above is not trivial, in that it has a purpose for both cognition and motivation and so is an example of a dual-purpose activity that we met in Chapter 6. By being fantastical, we can create scenarios that students won't have come across before and so we can be sure that students are transferring knowledge as opposed to simply recognising surface features that they have seen before. Students then will need to construct new meanings beyond those that have been explicitly taught in the introduce and practice phases of the lesson, where they will have the opportunity to make and break connections between different scientific ideas.

Of course, there is the risk that this type of activity may introduce some troublesome misconceptions with some students leaving the lesson looking for Green Children. This is certainly possible, but there are two responses that, I hope, will alleviate some of your anxiety. The first is that this scenario is not so fantastical after all. The mitochondria present in our own cells were once a free-living aerobic bacterium that got incorporated into a larger cell some 1.45 billion years ago and plants acquired their chloroplasts from a cyanobacterium-like cell in a similarly dramatic event. There is also the wonderful sea slug, *Elysia chlorotica*, that can live as a plant when given water, light and air, thanks to the algal prey it consumes that are packed full of photosynthetic machinery (Rumpho *et al.*, 2011). Second, maybe we shouldn't be so worried about mistakes after all, at least not in the short term. Every science teacher tells students that plants respire and photosynthesise, yet still many students think plants don't respire. Simply sticking to the facts doesn't seem to be working. Rather, we need both – that is, to tell the science story but then create opportunities to challenge and explore what students are thinking. Carefully planned confusion like this gets students to reflect and deliberate on what they know, allows teachers to see openly what students do and do not understand which, if coupled with feedback, can play an important role in developing understanding (D'Mello *et al.*, 2014).

Making connections

Maintenant, quand je dis que la confusion est une bonne chose, je parle du bon type de confusion. Not all confusion though, as you've just experienced, is a good thing! Good confusion, like a hook, focuses attention on the intrinsically important areas of the subject in a way that allows students to reassess what they think they know. Another way to do this is to ask students to connect two seemingly unrelated, but familiar, objects together using scientific processes (Figure 14.4). If students can explain how a roast chicken and an onion are connected by carbon atoms, this is good evidence

> **Instructions.** Work as a group to show, as many ways as possible, how a carbon atom present in a cooked roast chicken could become part of the cell wall of an onion cell. Add equations where you can.

Figure 14.4 Connecting familiar objects using scientific ideas

Source: Drawings by Arabella Green.

they understand the carbon cycle, and are not simply mimicking what they have been told. This type of task also involves retrieving information which, as we've seen in Chapter 10, plays an important role in memory.

As students solve this problem, they are integrating the scientific knowledge with aspects of their everyday life. In this case, a roast chicken and an onion are being connected through processes such as decomposition (chicken carcass is broken down by bacteria which release CO_2 from respiration), photosynthesis (CO_2 is taken in by plants and used to make sugar), translocation (sugar is transported from the leaf, through the phloem to the cells in the root) and cell wall synthesis. Alternative ideas may include burning the chicken, or indeed eating it and then considering what happens to the digested products that travel into the blood or out of the bottom. No carbon will be taken up directly from the soil by the plants though, as many students commonly believe.

This task can be adapted for many other topics and developed further by asking students to take the information from their answer sheets and then assemble the ideas together to write an extended answer. This layered approach – that is, to do the thinking in pairs first, get feedback and then take these ideas to work individually – is a good model you can use to structure any activity.

Making and testing things

With the pressures of examinations and curriculum content to get through, it can be easy to forget that science is a subject dedicated to explaining the natural world – and a big portion of this natural world involves living things that students have a right to experience. This could involve growing bacteria, making bread, exploring what coloured spaghetti birds prefer, or carrying out genetic crosses using *Drosophila*. The opportunities are endless. Plants represent a particularly attractive group of organisms to work with as they don't have feelings (as far as we know) and are relatively inexpensive.

The Science and Plants for Schools website (https://www.saps.org.uk/) provides a rich collection of practical activities involving plants. One such activity involves students following a standard procedure to make their own fertiliser which they then test on an unsuspecting seedling. Not only does this type of activity draw together two of the most enjoyable aspects of science – that is, making and testing things – it also unites ideas and skills from chemistry, biology and the nature of science. The websites below provide further practical resources that you can mine.

Practical biology – https://pbiol.rsb.org.uk/
Practical physics – https://spark.iop.org/practical-physics

Practical chemistry – https://edu.rsc.org/resources/collections
   /nuffield-practical-collection

Inquiring about socio-scientific issues

Integration, that is taking scientific knowledge and combining it with other aspects of our lives, is an important goal of learning science. By doing this, students come to see science as a body of knowledge that can be used to inquire and make sense of their lives as opposed to seeing science as a bunch of ideas that are confined to schools and science labs.

The empirical nature of science means that many questions that are of interest to young people can be explored practically through scientific inquiry. This could be an investigation lasting one or two lessons, or it could be a more extended project that lasts for a longer period of time. Often these types of question involve socio-scientific issues (SSIs) (see, e.g. Levinson, 2018). SSIs are so called because they consider the consequences of scientific knowledge to broader society which often generates controversy.

An example of a SSI might look at food wastage and packaging – a popular topic of conversation today. Plastic is getting quite a bad rap at the moment, yet packaging plays an important role throughout the world in reducing food wastage. Wrapping a cucumber in a thin plastic film apparently extends its life dramatically. It seems then that there is a genuine inquiry to be had in weighing up, using student gathered primary and secondary data, the relative advantages and disadvantages of packaging and this type of inquiry seems relevant to the lives of students at school and wider society (Table 14.1).

Table 14.1 Example of a scaffolded inquiry

Inquiry question	Should plastic packaging be removed from supermarkets?
How to organise?	How will you make sure everyone's ideas are heard? What do my friends think we should do? What is the best way to go about this?
Things to think about	Which vegetables or fruit will you investigate? What measurements are you taking? What equipment will you need? When are you taking your measurements? Sketch what your graph(s) might look like.
Collecting the data	How will you record the data? How will you make sure your data are accurate? How are you managing risk and keeping safe?
Interpretation	What does the data tell you? What can you do about it?

Source: Adapted from Levinson (2018).

Interest in inquiries like this could be stimulated through a news article or by asking students to bring in examples of supermarket packaging from home. A concept cartoon (Keogh and Naylor, 1999), showing the different perspectives of farmers, consumers and supermarkets could help clarify the controversy before asking students to raise their own questions to investigate. Depending on the question, students may collect primary or secondary data, for example, investigating how wrapping small amounts of food in different materials affects the rate of decomposition or doing some research into food packaging and food spoilage. It's a good idea to use sealable bags here to prevent breathing in any nasty microbes, although I do get the irony of using a plastic bag to investigate the overuse of plastics. Finally, students analyse their results and make tentative conclusions that might result in some action being taken. Depending on the time available, students may want to present their findings in the format of a scientific poster – just as scientists would do for a conference.

Inquiries like this not only have a role in helping students to experience what the practice of science feels like that we discussed in Chapter 3, they also play an important role in getting students to engage with personal and social aspects that deepens knowledge – 'making us more critical, practical and understanding' (Levinson, 2018, p. 31). This deepening of understanding happens as students come to see the same problem from different perspectives. In this case, the issues of packaging from the perspectives of the supermarket, the farmer, the packaging company and the consumer.

Challenges

Trying to describe a non-example here is probably not going to be that useful, so I won't. Instead, thinking through some of the challenges involved when we ask students to apply and integrate their knowledge might be more useful. The first challenge concerns prior knowledge. If students have insufficient prior knowledge, then they won't be able to reason about the problems in beneficial and deep ways. This means we need to plan carefully what knowledge and skills students will need for application tasks and build these into the curriculum, before students are asked to deploy them. Because application tasks are complex, students may need access to resources such as peers, books, videos or websites during these activities. Perhaps, though, the greatest resource students need is time as they need to make mistakes, struggle and resolve any confusion (Land *et al.*, 2014). If time is in short supply, we will leave students with no choice but to think superficially about the scientific ideas.

Taking it further

- Students carry out a project, lasting a number of lessons, to investigate a question that matters to them – this could involve some organisms, for example, woodlice or plants or going outside to collect data.
- Extended writing – students write an essay to an open question, for example, explaining why big fierce creatures are rare. This gives students the opportunity to bring together lots of related concepts from different topic areas. To make sure this is successful, spend time planning out the essay first, possibly using concept maps and then take these ideas to create a more linear plan.
- Students spend time creating their own models to explain a scientific idea, for example, why food chains don't go on forever. They then present their model to peers who identify strengths and limitations using some pre-identified criteria.

Summary

Learning science requires students to reason about the natural world using both everyday experiences and scientific ideas. Whilst a substantial amount of time in classrooms needs to support students to acquire the scientific ideas in carefully guided ways, students also need time to use their scientific knowledge to solve problems, create and discover. Not only is this motivating, because students can see what the scientific ideas enable them to do, it also creates a time for students to reflect upon and refine their own understanding in more subtle ways than is possible when all students receive similar instruction. The hope is that these periods of application and integration in the classroom encourage students to take the scientific ideas and make them their own, meaning that they use these ideas beyond the school gates, either to inquire about something or to do something differently.

Bibliography

D'Mello, S., Lehman, B., Pekrun, R. and Graesser, A. 2014. Confusion can be beneficial for learning. *Learning and Instruction*, 29, pp. 153–170.
Green, J. B. n.d. *The green child*. Available at: https://thescienceteacher.co.uk/photosynthesis/. [Accessed: 15 December 2019.]

Hershey, D. R. 2004. *Avoid misconceptions when teaching about plants*. Available at: https://www.thevespiary.org/library/Files_Uploaded_by_Users/llamabox /Botany/Avoid%20Misconceptions%20When%20Teaching%20about%20 Plants%20by%20David%20R.%20H...pdf. [Accessed: 15 December 2008.]

Kalyuga, S., Chandler, P. and Sweller, J. 1998. Levels of expertise and instructional design. *Human Factors*, 40(1), pp. 1–17.

Keogh, B. and Naylor, S. 1999. Concept cartoons, teaching and learning in science: An evaluation. *International Journal of Science Education*, 21(4), pp. 431–446.

Land, R., Rattray, J. and Vivian, P. 2014. Learning in the liminal space: A semiotic approach to threshold concepts. *Higher Education*, 67(2), pp. 199–217.

Levinson, R. 2018. Introducing socio-scientific inquiry-based learning (SSIBL). *School Science Review*, 100(371), pp. 31–35.

Rumpho, M. E., Pelletreau, K. N., Moustafa, A. and Bhattacharya, D. 2011. The making of a photosynthetic animal. *Journal of Experimental Biology*, 214(2), 303–311.

Section 5
Responsive teaching

Making thinking visible so feedback can take place

Powerful idea focus

- *The diversity of organisms, living and extinct, is the result of evolution by natural selection.*

Pedagogy focus

- looking at ways to reveal what students are thinking about so that feedback can take place.

To get better at something, we need feedback. Feedback is simply information about our current performance or understanding that we can then use to confirm, reject or change what we know, so we can get better.

When we're learning how to play most sports, we get some feedback from the task itself – either the ball goes over the net or into the hole. When we're cooking, we get feedback from how the food smells and tastes. If we smell burning, we turn down the gas or flip the food over, and if it's too spicy, we add less chilli next time. There are, though, no acrid smells or nets to get over when learning science. This leaves both teachers and students in a tricky situation as, for the most part, there is little feedback available when learning scientific ideas as thinking is invisible and many wrong ideas, as we saw in Chapter 4, still work. This means we must devise goals, or metaphorical holes, using questions that we can use to gauge how well we are doing. I use the word 'we' here not to refer to you and me, but to reflect the fact that feedback is for both students *and* teachers.

▌Feedback

Before we look at feedback in more detail, it is important to recognise that not all feedback is a good thing. In a large study, exploring a number of research papers, Kluger and DeNisi (1996) showed that one third of feedback actually decreased performance. This slightly unnerving finding may make more sense if we stop and reflect on our own experiences of receiving feedback. How motivating was it? Was it so general it was difficult to know what to do with it? Or, at the other extreme, perhaps feedback was so specific that it only helped to improve that particular sentence that we never used again?

To avoid some of these problems, the focus of this chapter is on feedback that focuses improvement on understanding the subject. We will first look at ways to create tasks that reveal student thinking and then consider how students can get feedback that will improve not just their performance on that task but their understanding of the subject more generally.

▌The importance of goals

Often the goals used to track progress in lessons are examination questions. So, if we want to see if students have understood a concept, we may give them an examination question at the end of the lesson to answer. There is certainly value in getting students to practice questions that in turn allow them to become familiar with the idiosyncrasies of awarding bodies, made clear through studying mark schemes and examiner reports. The problem with this approach, though, is that examination questions were written to grade students, they were not designed to provide feedback on reasoning – in other words, examination questions tell us what students got right and wrong, but they don't tell us *why* students got them right or wrong. It is understanding this reasoning that is important in science classrooms because it guides next steps, for teachers and students, of what to do to improve. This means that we need to create tasks that provide more formative information and that these tasks may look very different from the end goal (Christodoulou, 2016).

To explain what I mean, let's consider how we can explore what students know when teaching about the final powerful idea of this book, that is, *the diversity of organisms, living and extinct, is the result of evolution by natural selection.*

The story of the long neck

Giraffes, more properly known by their binomial name of *Giraffa camelopardalis*, are the world's tallest mammals and form an ideal context to study evolution. Their necks average 1.8 m in length and have a mass of about 270 kilograms – that's four times the mass of an average human (whatever that means). Despite this enormous length, giraffe necks have exactly the same number of cervical vertebrae as us – seven – although admittedly their vertebrae are longer. Their legs are long too – 1.8 m – as are their 46 cm tongues, which they use to carefully strip leaves from the top of the rather thorny acacia trees they feed from.

Giraffes evolved their long neck according to the theory of evolution by natural selection, proposed by Charles Darwin in 1859. As we saw in Chapter 3, the word theory here does not mean evolution is a good guess of what happened, rather there is a large body of accepted empirical evidence that supports this idea. At the core of Darwin's theory is the idea that all organisms are related and descended from a common ancestor. Giraffes share a common ancestor with us about 96 million years ago (http://www.timetree.org/). Their closest living relative is the slightly odd-looking okapi that is native to the Democratic Republic of Congo. Okapi, also known as the forest giraffe, have stripes on their legs reminiscent of a zebra and stand at a slightly disappointing 1.5 m tall relative to their long-necked cousins. They belong to the same family as giraffes – the Giraffidiae (Table 15.1), and both giraffes and okapi belong to the same class as us – Mammalia. This system of classifying organisms then not only is helpful in grouping organisms that might share similar and useful properties that are beneficial for humans (see, for example, the story of Taxol to treat breast cancer in Bryan, 2011), it also reflects the evolutionary relationships between organisms.

Table 15.1 The classification system groups together organisms that are related

	Forest giraffe	*Giraffe*	*Human*
Kingdom	Animalia	Animalia	Animalia
Phylum	Chordata	Chordata	Chordata
Class	Mammalia	Mammalia	Mammalia
Order	Artiodactyla	Artiodactyla	Primates
Family	Giraffidiae	Giraffidiae	Hominidae
Genus	*Okapia*	*Giraffa*	*Homo*
Species	*O. johnstoni*	*G. camelopardalis*	*H. sapiens*

▦ Why the long neck?

Students might explain how giraffes came to acquire such a long neck in many different ways. They might argue giraffes were put on Earth in their long-necked form by a divine creator. This argument is hard to justify when you consider the number of similarities between giraffes and other organisms in both their biochemistry (all organisms use the same genetic code) and in their comparative anatomy – giraffes and other mammals share a number of structures, for example have the same number of neck vertebrae. If giraffes were 'designed' from scratch, it seems unlikely the designer would be constrained by the anatomy of mammals and the biochemistry of life.

A second hypothesis might suggest that giraffes acquired long necks as a consequence of gradual stretching. As giraffes struggled to reach the leaves at the top of the acacia trees they stretched various structures throughout their lifetime to get there – just like a body builder might develop bigger muscles as they lift larger weights (sorry, masses). These 'stretched' structures are then seen to be passed on to offspring. A third explanation, and definitely my favourite, is that long necks evolved as a result of repeated natural selection – that is, those individuals in the population with slightly longer necks produced more offspring than those with shorter necks. Because neck length was controlled by genes, this trait was inherited by the offspring. This process was repeated over many generations giving rise to the giraffes we see today.

All of these hypotheses 'work' in that they explain how giraffes could have evolved long necks. They just differ in how much empirical evidence there is available in support of them. The challenge then for science teachers who are teaching about evolution, and indeed pupils learning these ideas, is to discern between the scientific idea and the non-scientific ones. This is necessary because we'll need to go looking for all of these ideas when trying to work out what students are thinking. This means designing tasks that reveal both what ideas students hold and why they hold them so that appropriate feedback can take place. Not only does this feedback need to help students to evaluate their own ideas, it also needs to give them guidance on how to improve.

▦ Assessment for learning

This is the idea of assessment for learning (AfL) (Black and Harrison, 2004), sometimes referred to as responsive teaching. The starting point for AfL is to recognise the main ideas that students need to know – these are the goals that we framed using key questions in Chapter 8. Then, we want to find ways in which we can take these goals and transform them into an activity that allows teachers and students to see when they have been achieved. These types of activities (see e.g. Mintzes *et al.*, 2001) could take place at various points throughout the lesson. So, rather than seeing AfL as some standalone activity, see it as part of the learning process itself.

Defining the assessment range

To understand how giraffes acquired a long neck, students will need to piece together the six ideas below to construct an overall understanding of how populations, and not individuals, evolve by natural selection over time.

1. There is a struggle to survive and reproduce as **resources are limited**.
2. Organisms show **variation** in characteristics that may affect their success of surviving.
3. Parents with traits that allow them to survive longer and reproduce more will **produce more offspring** than the other individuals in the population.
4. If these beneficial traits, or adaptations, are caused by genes rather than the environment, then the offspring will **inherit** them.
5. Over time, the **proportion** of individuals in the population with the adaptation will increase.
6. If two populations of one species become so different from each other that they can no longer interbreed successfully, then two **species** have been formed.

Now that we have identified what we want students to learn, we can think about the range we are interested in assessing (Figure 15.1).

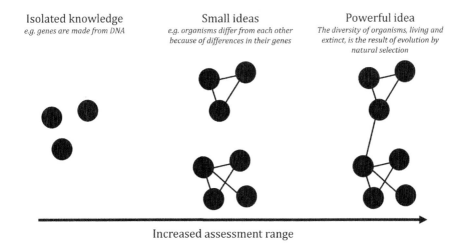

Figure 15.1 Thinking about the assessment range matters when asking questions

In the beginning, when students are first learning about any scientific idea, this might involve checking to see if they have learnt bits of isolated knowledge, for example, that they understand what a population is or know the name of the ship that Darwin sailed on. Later, we might want to know if students have acquired an understanding of smaller ideas such as variation or adaptation. Finally, we will want to see if students can piece these smaller ideas together to acquire an understanding of the powerful idea itself.

Whatever the range of assessment, whether it's facts, smaller ideas or larger ones, it generally starts with a good question. These questions then act as mini-goals for students to answer and so reveal their thinking, allowing feedback to take place. What we do with a question is the interesting bit. We might use it to structure a discussion, create a multiple-choice question or provide a stimulus for an investigation or project. Questions then are the starting point to find out what students know and we use these questions to create tasks whose purpose is generally fourfold:

- To draw out and extend ideas.
- To display thinking to others.
- To support feedback.
- To enable feedback to be used.

Let's now use some actual examples to explore what these tasks might look like. At the end of this chapter, we will then think about how to respond.

Teacher questioning

Teacher questioning looks to elicit and extend students' ideas so that self, peer and teacher evaluations and feedback are possible.

Chin (2007) explored questioning in 36 science lessons and identified a number of different approaches that teachers use. I have taken three important ones below to illustrate how questioning approaches differ but there are many more you can explore by reading the paper. In each case, questioning can be seen as a tool to both draw out and evaluate what students know but is also a tool to promote learning by scaffolding student discussions.

- Socratic questioning – using questions to prompt, guide and elaborate student thinking; for example, you said humans evolved from chimpanzees, if this is the case, then why are chimpanzees still here?
- Pumping – putting the onus on the student to provide more information; for example, can you give an example of that?

- Focusing and zooming – use questions to move between the macroscopic, cellular and submicroscopic levels; for example, can you explain how a gene mutation might lead to a bacterium becoming resistant to penicillin?

Questioning tends to go awry in classrooms when it starts 'fishing for an answer'. This fishing looks a bit like this (Scott *et al.*, 2006):

Teacher question → Student A can't answer → Teacher question → Student B can't answer → Teacher question → Student C can't answer

Instead, a richer dialogue looks to develop a chain where teacher feedback is used by the student to create a further response:

Teacher question → Student A can't answer → Teacher feedback → Student A responds → Teacher feedback → Student B adds to idea → Teacher feedback → Student B

This type of chain questioning is challenging to do and requires strong subject knowledge and good behaviour in the classroom. It requires, too, the ability to adapt the lesson to take account of students' ideas, recognising that the ultimate goal is for students to be able to take what they know already about an idea and connect it to the scientific way of thinking. This requires us to decide whether to close down discussions and simply tell students when they get a question wrong or whether to open up discussions to explore student thinking. There is no easy rule here to follow, other than to recognise that opening up dialogue comes at a cost as it slows pace and often temporarily confuses. The question, then, is, are the ideas we are discussing worth spending time on? For me, the answer is yes when they are conceptually tricky, such as the idea between laws and theories, and no when they are not, such as the name of the ship that Darwin sailed on.

Concept cartoons

Whilst teacher questioning is a useful assessment tool, it often involves only a few students. We can overcome this problem by structuring discussions so that all students have time to discuss their ideas. This discussion helps students to develop their ideas through justification and provides an opportunity for feedback – both from their peers and the teacher. The challenge then is how to create a discussion task that gets students talking about what you want them to talk about and that you can see, or at least hear, what students are thinking.

Concept cartoons (Keogh and Naylor, 1999) are one way in which we can help students take part in focused discussions. A concept cartoon is a cartoon that presents different ideas and asks students to come up with what

Figure 15.2 A cartoon to explore what students think are the causes of evolution

Source: Drawing by Arabella Green.

they think is the correct idea. You can include misconceptions here. I have attempted an example in Figure 15.2 but there are many others available.

Discussions like this can be structured further by providing scaffolds to prompt thinking. For example, students can be asked to write down who they agree with and why and who they disagree with and why before coming up with what they think (Chin and Teou, 2010). By moving around the room and listening to conversations, it's possible to not only hear what students think is the right idea but also understand some of the reasoning behind their thinking. This type of information is dynamite when planning next steps for feedback.

Drawing pictures

With any discussion, there is always the danger that one or two students will dominate. Whilst this problem can be reduced by assigning numbers to students and then rotating around the group, there are times when we may want students to commit pen to paper. The problem with this approach is that it can be very difficult to know what students are writing about because writing is very small. Pictures, however, are much easier to see at a distance and can provide an important window into students' thinking (Ainsworth *et al.*, 2011). Let me use an example.

Imagine we wanted to find out if students understand that evolution involves the descent of organisms from a common ancestor – that is, the

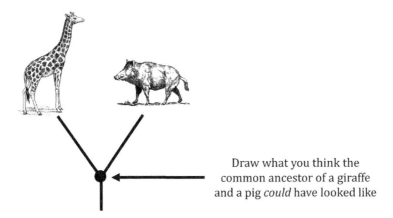

Draw what you think the common ancestor of a giraffe and a pig *could* have looked like

Figure 15.3 Drawing pictures can be a useful way to find out what students are thinking; this approach can be applied to lots of topics

Source: Images taken from https://pixabay.com/.

most recent individual from which two organisms are descended. We could ask them to write this down but then we would be stuck in how to review this task. Instead, a better approach might ask students to draw what they think the common ancestor of a pig and giraffe might have looked like (Figure 15.3). If students draw an animal that is half giraffe and half pig they don't understand evolution – that is, organisms descend from a common ancestor with modification. If, however, students are able to recognise that the common ancestor was likely to have features common to both species, that is, they draw something with four legs, a trail and hooves, then this is evidence that they understand what a common ancestor represents. If they can then argue that the ancestor may have had a long neck and that this was lost during the evolution of pigs then even better (although this hypothesis is very unlikely considering related species also have short necks – a nice opportunity to discuss the idea of parsimony in science, explanations with the fewest steps are favoured).

Sketching graphs

Another type of picture you can ask students to draw is a graph. As we've seen throughout this book, understanding science is all about making connections. These might be connections between cause and effect, for example knowing that ultraviolet radiation increases the risk of genetic mutations, or the connection might be between structure and function, in knowing that giraffes with longer necks can reach the leaves at the top of the trees. This 'connectedness' means that graphs are a great way to explore what students know because a graph is simply a picture that shows the relationship between two quantities.

At the heart of understanding evolution by natural selection is knowing that there must be differences between organisms within a population to begin with. If all organisms are identical, then natural selection has nothing to work on as everything is equally adapted. We call these differences between organisms variation. This variation could be between members of the same species, for example, eye colour in humans or variation between individuals of different species, for example, giraffes have a black tongue, thought to protect them from sun burn, whereas humans have a red tongue. This type of variation can be caused by changes in our DNA or by the environment, but more often than not it is caused by a combination of both. Natural selection can only act on variation that is caused by changes in our genetic material because only these differences are inherited.

To explore what students know about variation, we might ask them to sketch a graph showing the frequency of a specific trait such as neck length (Figure 15.4). This graph can then provide the stimulus for a discussion. Not only are graphs easy for us to see and so we can evaluate what students know, they are also a good way of focusing students' attention on one or two variables at one time. This type of activity can be modified too, for example, by asking students to sketch two lines to show a population when food is plentiful and the same population a number of generations later after a drought.

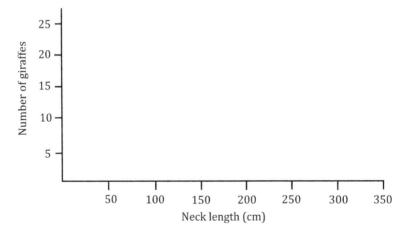

Figure 15.4 Getting students to sketch graphs is a quick way to see what they think

Multiple-choice questions

Multiple-choice questions are another quick way to find out what students are thinking and can be structured to directly explore misconceptions or common errors identified previously in marking. There are many excellent examples of multiple-choice questions in the public domain that you can use (see, for example, https://www.stem.org.uk/best-evidence-science-teaching). You can also have a go at designing these questions for yourself by drawing on published research on common misconceptions and using the guidance below.

In general, a multiple-choice question consists of a stem that clarifies the question and a series of possible answers that includes the right or 'best answer' along with some distractors (Figure 15.5). Distractors are plausible wrong answers and need to be chosen carefully so that students are not just picking the right answer because the distractor is so far-fetched they are left with no other choice! Using three or four plausible wrong answers is a good general rule to follow as it means students have to process the information, yet can still be successful if they have the necessary knowledge (Butler, 2017).

Figure 15.5 shows an example of a two-tier multiple-choice item, so called because it requires students to also give a reason for their answer. Here, the distractors (answers a and c) are making sure students understand that relatedness is based on how recently two species shared a common ancestor. Trees like this can be hard for students to interpret, often because they make

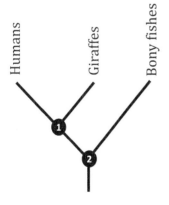

Circle the letter corresponding to the correct statement about the picture. Then provide a reason.

a) Giraffes are more closely related to bony fishes than to humans

b) Giraffes are more closely related to humans than to bony fishes

c) Giraffes, humans and bony fishes are all equally related

d) It's impossible to tell

Reason:

Figure 15.5 A multiple-choice question; the correct answer is (b) because humans share a common ancestor (1) with giraffes more recently than they do with bony fishes (2)

it seem that humans are somehow 'higher' than other organisms. This is of course not true; humans are no more higher than giraffes or any species on Earth, including bacteria — our DNA has been evolving for the same amount of time as everyone else's. Rather, humans have gone down an evolutionary path of big brains as opposed to big necks or streamlined bodies.

Responding

So far we have focused on ways to elicit students' ideas, but we haven't really considered how feedback then takes place. Whenever we are talking about feedback in lessons, we are talking about the flow of information that results in improved understanding or knowing how to do something. Figure 15.6 attempts to summarise how this flow of information might work by identifying some of the main sources of student feedback during a task; that is, feedback from peers, the teacher or directly from the task itself.

We will now explore each of these feedback channels in a little more detail.

Peer feedback: from student to student

Learning is a social process and so when students carry out a task together they learn from each other as they have to justify and clarify their reasoning. This often takes place in short-paired discussions but it can also happen when students come to the teacher's board at the front of their classroom and display their thinking as they solve a problem. We looked at how this

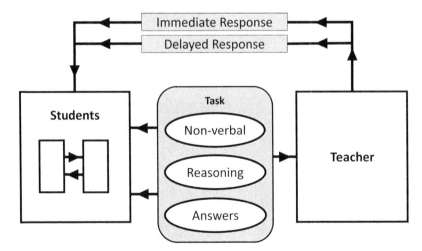

Figure 15.6 Ways students and teachers receive feedback from a task; feedback is represented by lines with arrows

might work in Chapter 10 to review one or two questions from the rewind phase. Peer feedback can also take place when a written task or product is reviewed. To make this as successful as possible, use clear success criteria that students use to assess the work against, model what good peer feedback looks like first and give sufficient time for students to then act on their feedback, perhaps through redrafting their answer or using this learning to answer a related question.

Oral teacher feedback

To provide feedback, teachers must first notice and interpret what they see (Cowie *et al.*, 2018). This noticing will involve listening to students' answers and watching for other non-verbal cues, such as wrinkled brows, that might signal confusion. Once we have this information, we must then decide whether to respond to misunderstandings immediately or in subsequent lessons (Figure 15.6). If the error is easily addressed, corrective feedback using teacher explanations and the whiteboard can be effective. If, however, the misunderstanding revealed is more complex than a simple error, we may want to delay feedback until the next lesson when there has been time to plan how to address it.

A slightly different strategy to corrective feedback is to correct the misunderstanding by presenting some additional information that shows students why their original thinking might be wrong, or at least insufficient. For example, if students think that giraffes evolved a long neck through a Lamarckian, use and disuse hypothesis, then we might ask them if children born to parents with tattoos also have tattoos, and then ask them to explain why not. This is using a questioning approach to refine students' understanding. This type of activity can be done by posing a question and then allowing time for a quick paired discussion, as it gives students time to explore and revise their thinking first as opposed to simply accept the teacher's ideas without much thought as to why their ideas were wrong in the first place.

Written teacher feedback

Marking student work can be incredibly valuable in getting individualised feedback from students, but it's also time consuming, so plan carefully what you intend to mark. As you read student work, it's helpful to make a note of the common errors that crop up on a blank copy of the task. Where possible, add your written feedback in the body of the student work next to the error, so students see what the feedback relates to. If students have made careless mistakes simply mark it as incorrect without giving the right answer (Elliott *et al.*, 2016). Students find comments helpful if they are posed as questions that might challenge wrong ideas or extend thinking as this makes more sense. Avoid simply posing stretch questions that don't address errors made in the work. Once finished marking, you can then set about thinking how you will address common errors in the next lesson.

Focus on the errors that are most important for understanding the subject, rather than addressing errors that are specific to the task. For example, it's more important that students understand variation than how many vertebrae giraffes have. Some errors, such as spelling, can easily be corrected, perhaps at the start of a lesson in a rewind activity. Other errors, often conceptually tricky, may require longer to unpick and so will need an activity, such as a concept cartoon, demonstration or teacher explanation that addresses the error. This may involve presenting the mistake alongside the correct idea, so students can see why right is right and why wrong is wrong. Be wary of only showing students' mark schemes – these documents were written for experts to use and were not written as teaching tools for beginners. If you're not careful, students will correct their answers simply by transferring words from the mark scheme to their work. Instead, get students to use the feedback to complete another, similar task to prove they understand and so create time in lessons for pupils to respond to feedback.

Task-mediated feedback

It might sound strange but the task itself can provide feedback too. This is easier to understand if we think about a practical task. In chemistry, one of my favourite challenges, created by Sana Badri, is to provide students with a range of aqueous solutions ($CuSO_4$, NaOH, $BaCl_2$) and common lab apparatus and to ask them to make a white precipitate floating in a colourless solution. When students get the right answer, they and the teacher get immediate feedback. Why? Because everyone can see for themselves a white precipitate floating in a colourless solution. Or, if there's an error, a white precipitate in a blue solution; in which case there is an excess of copper ions floating about and so adding a little more sodium hydroxide next time should do the trick.

Summary

Despite years of science education, many students leave school with a poor understanding of both the nature of science and the scientific ideas themselves. This means that students are sitting in lessons where they understand very little. Yet, visit many science classrooms and you will be forgiven for thinking that on the whole, things are going swimmingly. The problem is that students are particularly adept at hiding what they don't know and we often confuse clear teacher explanations as evidence of learning. This means we need to devise tasks that reveal honestly what students are thinking and that also provide some information as to why they are thinking it. In this chapter, we have looked at how drawing, multiple-choice questions, concept cartoons, practical work and teacher questioning all help to reveal what students are thinking about. We have then considered how this information can be used to provide feedback that focuses on improving performance in the subject, as opposed to thinking only about improving performance on the task. Whilst tasks such as these can be used in a standalone way, they are most effective when seen to form part of the learning process itself.

Bibliography

Ainsworth, S., Prain, V. and Tytler, R. 2011. Drawing to learn in science. *Science*, 333(6046), pp. 1096–1097.

Black, P. and Harrison, C. 2004. *Science inside the black box: Assessment for learning in the science classroom*. London: NfER Nelson.

Bryan, J. 2011. How bark from the Pacific yew tree improved the treatment of breast cancer. *The Pharmaceutical Journal*, 287, p. 369.

Butler, A. 2017. *Multiple-choice testing: Are the best practices for assessment also good for learning?* Available at: https://www.learningscientists.org /blog/2017/10/10-1. [Accessed: 1 January 2020.]

Chin, C. 2007. Teacher questioning in science classrooms: Approaches that stimulate productive thinking. *Journal of Research in Science Teaching: The Official Journal of the National Association for Research in Science Teaching*, 44(6), pp. 815–843.

Chin, C. and Teou, L. Y. 2010. Formative assessment: Using concept cartoon, pupils' drawings, and group discussions to tackle children's ideas about biological inheritance. *Journal of Biological Education*, 44(3), pp. 108–115.

Christodoulou, D. 2016. *Making good progress? The future of assessment for learning*. Oxford: Oxford University Press.

Cowie, B., Harrison, C. and Willis, J. 2018. Supporting teacher responsiveness in assessment for learning through disciplined noticing. *The Curriculum Journal*, 29(4), pp. 464–478.

Elliott, V., Baird, J. A., Hopfenbeck, T. N., Ingram, J., Thompson, I., Usher, N., Zantout, M., Richardson, J. and Coleman, R. 2016. *A marked improvement? A review of the evidence on written marking*. London: Education Endowment Foundation.

Keogh, B. and Naylor, S. 1999. Concept cartoons, teaching and learning in science: An evaluation. *International Journal of Science Education*, 21(4), pp. 431–446.

Kluger, A. N. and DeNisi, A. 1996. The effects of feedback interventions on performance: A historical review, a meta-analysis, and a preliminary feedback intervention theory. *Psychological Bulletin*, 119(2), pp. 254–284.

Mintzes, J. J., Wandersee, J. H. and Novak, J. D. 2001. Assessing understanding in biology. *Journal of Biological Education*, 35(3), pp. 118–124.

Scott, P. H., Mortimer, E. F. and Aguiar, O. G. 2006. The tension between authoritative and dialogic discourse: A fundamental characteristic of meaning making interactions in high school science lessons. *Science Education*, 90(4), pp. 605–631.

Conclusion
Time to reflect

I began this book by asking you to hold on tight as we embarked upon a journey into the powerful ideas of science and how to teach them. After 15 chapters, we are now approaching the end of this journey together and so it feels like an appropriate time to reflect, draw ideas together and tie up any loose ends.

Reflection one: defining aims matters

My first reflection is more of a realisation. The realisation that unless the aims of school science are clarified, any attempt to evaluate the effectiveness of a specific pedagogy, lesson plan or cognitive load effect is not possible. This means that without first specifying the aims of school science, claims of one pedagogy or curriculum model being better than another are not particularly helpful. The important question we should be asking ourselves is: better for what?

In this book, I have argued that a major aim of school science is for students to learn about 13 powerful ideas of science. Indeed, this was going to be the sole focus of this book before I started writing. However, as I read more and wrote more, I came to realise that an education in science is more than just learning about the conclusions of scientists, important though they are. It's also about understanding the nature of science itself. That is, as we saw in Chapter 3, to understand how scientific knowledge is different from other knowledge, how it is generated and what its implications are for society. To learn science at school means to learn about the scientific ideas *and* to understand the stories of how they came to be. This doesn't require a detailed history of how every scientific idea arose, that would get tedious. Rather, students need sufficient examples to appreciate scientific knowledge for what it really is – tentative, subjective, creative, empirical and embedded in a society that determines what gets measured. Not only is understanding the nature of science intellectually interesting, it also serves an important role in helping students to navigate scientific claims in the 'real world'.

Reflection two: scientific inquiry is integral to what science is

If we want students to understand how scientific knowledge is generated, we need to create opportunities for them to engage in scientific inquiries. These are moments when students investigate questions that matter to them and there is some choice in how to do it, in a similar way to real scientists. This is not to say that students should be expected to stumble upon important ideas such as evolution by natural selection, the chromosome theory of inheritance or how to perform distillation. Neither does this mean every lesson needs to be a scientific inquiry. Rather, there should be dedicated time in the curriculum where students bring all of their knowledge and skills together to inquire about something. Not only are these experiences incredibly motivating (for teachers and students), they also induct students into the inner workings and complexities of scientific practices and so play an important role in demystifying science, making it feel relevant and not just something that other people do.

Reflection three: powerful ideas change how we look at the everyday world

Often in science lessons we start with a 'bang' or some other curious demonstration to trigger interest before moving on to 'the science'. However, the real power of scientific ideas lies not in their ability to explain the unusual, already interesting moments, but their ability to transform our bland everyday experiences into moments of awe and wonder. So, whilst an ice cube floating on water is not that interesting, show me a solid floating on its liquid form and now I'm excited!

Reflection four: knowledge needs to be portrayed in many different ways

In Chapter 2, we saw how powerful ideas emerge when lots of knowledge, bounded in concepts, are connected together. The central role played by knowledge in enabling students to understand scientific ideas means lesson planning starts with defining what it is we want students to learn and why. This should then be informed by a carefully sequenced curriculum that plans for progression over time. However, as we saw in Chapter 8, knowledge is more than just information; it needs to be portrayed in different ways so that students come to understand scientific ideas in all of their dimensions. Cells are not just flat two-dimensional drawings on a page or a definition – they need to be visualised in three dimensions using models, be seen to move under a microscope and appreciated for how incredible they are – after all, they form you. Whether these portrayals are through demonstrations, dissections, textbooks, videos, models or teacher explanations, they serve as objects that teachers *and* students use to come to understand the knowledge and so are as critical to learning science as the definitions themselves.

Reflection five: don't overly constrain assessment

Throughout this book I have made the case that knowledge plays a central role in understanding science. However, as we saw in Chapters 4 and 5, ideas are learnt through a personal process of construction, they are not learnt *de novo* through transmission. So whilst careful teacher explanations play an important role in helping students to understand scientific ideas, they are unlikely to be sufficient, meaning that many students construct ideas that are different, often subtly, from the scientific way of thinking. This means that, as we saw in Chapter 15, we need to devise activities that reveal what students are thinking about and why. If we overly constrain these tasks, for example, only using closed questions or gap fills, we may inadvertently assume students have understood, when in fact they haven't had the opportunity to make any mistakes, rendering feedback impossible.

Reflection six: plan lessons thinking about both cognitive *and* affective processes

Cognitive load theory, that we met in Chapter 5, provides a useful model to inform how we design instructional materials. However, it tells us less about *why* students want to learn these ideas in the first place. Planning science lessons needs to consider *both* affective factors, such as interest, emotion and motivation, as well as cognitive factors, such as the limited capacity of working memory, when dealing with new information. Of course, as we saw in Chapter 6, there is an important relationship between the two and that success and autonomy – two important sources of motivation – are dependent on knowing lots of stuff. Equally, as we saw in Chapter 11, triggering interest in science should look to use the intrinsic aspects of the subject and not distracting details that may get students thinking about the wrong things.

Reflection seven: not all practice is equal

Even meaningful moments can be forgotten, though, and so practice is an important part of the learning process by giving students the opportunity to move at different rates, make mistakes and refine what they know. But, as we saw in Chapter 13, this practice needs to be designed differently depending on whether students are practicing conceptual knowledge or procedural knowledge.

Reflection eight: you can do well in science without understanding it

Reflection eight is quite a frightening one. It is perfectly possible to do well in science examinations, at least in the short term, without understanding much science at all, as any high-stakes assessment can be gamed. Rote learning, explored briefly in Chapter 7, can be a remarkably effective strategy. This means that teachers and schools cannot rely on examination results alone as being the yardstick by which to measure success in science and we shouldn't rely on the assessment to set the standard. If we do, we will privilege tidy definitions and formula triangles, that are easy to reproduce, over activities that develop a deeper expert-like understanding. Instead, it falls to science teachers and the curriculum to set the standard of what it means to understand science.

Reflection nine: science is diverse and difficult to learn

Perhaps, though, my greatest reflection from writing this book is to recognise the complexities involved in learning (and teaching) science, or what Jonathan Osborne more accurately refers to as 'crazy ideas'. Not only are scientific ideas abstract, counterintuitive and exist at many different levels, learning them involves both learning and unlearning (or supressing) ideas too. And then there is the added challenge that science is made up of three very different subjects that themselves have their own rules and idiosyncrasies that students must navigate. This is why when we look for evidence of what works in science, we are probably best to focus our attention on the research that informs classroom practice at the topic level. That is, how to teach ionic bonding better or how to overcome misconceptions about current getting used up in circuits, rather than looking for evidence of what works across all subjects. If we simply search for general 'science' solutions, we probably won't get so far.

Reflection ten: teaching science is wonderful

So to my final reflection. It is because scientific ideas are radically different from everyday thinking that science teachers get to change how young people see themselves and the world around them in ways that few people can. This makes teaching science wonderful.

I wish you all the best as you take the ideas from this book and make them your own.

Lesson planning templates

Powerful idea(s):

Unit title:

Lesson title:

What is powerful about these ideas?

	WHAT?		HOW?	
	Key question	Expected answer [misconception]	Portrayed	Used
1				
2				
3				

Phase	Time (min)	What students are doing	What the teacher is doing	Key question
Rewind				
Transition				
Trigger interest and activate prior knowledge				
Transition				
Introduce				
Transition				
Practice to build understanding				
Transition				
Apply and integrate				
Transition				
Reflect and review				

Index

Page numbers in *italics* refer to figures and those in **bold** refer to tables.

Printed in Great Britain
by Amazon

79627926R00136